Springer Complexity

Springer Complexity is an interdisciplinary program publishing the best research and academic-level teaching on both fundamental and applied aspects of complex systems-cutting across all traditional disciplines of the natural and life sciences, engineering, economics, medicine, neuroscience, social and computer science.

Complex Systems are systems that comprise many interacting parts with the ability to generate a new quality of macroscopic collective behavior the manifestations of which are the spontaneous formation of distinctive temporal, spatial or functional structures. Models of such systems can be successfully mapped onto quite diverse "real-life" situations like the climate, the coherent emission of light from lasers, chemical reaction-diffusion systems, biological cellular networks, the dynamics of stock markets and of the internet, earthquake statistics and prediction, freeway traffic, the human brain, or the formation of opinions in social systems, to name just some of the popular applications.

Although their scope and methodologies overlap somewhat, one can distinguish the following main concepts and tools: self-organization, nonlinear dynamics, synergetics, turbulence, dynamical systems, catastrophes, instabilities, stochastic processes, chaos, graphs and networks, cellular automata, adaptive systems, genetic algorithms and computational intelligence.

The three major book publication platforms of the Springer Complexity program are the monograph series "Understanding Complex Systems" focusing on the various applications of complexity, the "Springer Series in Synergetics", which is devoted to the quantitative theoretical and methodological foundations, and the "Springer Briefs in Complexity" which are concise and topical working reports, case-studies, surveys, essays and lecture notes of relevance to the field. In addition to the books in these two core series, the program also incorporates individual titles ranging from textbooks to major reference works.

Editorial and Programme Advisory Board

Understanding Complex Systems

Founding Editor: S. Kelso

Future scientific and technological developments in many fields will necessarily depend upon coming to grips with complex systems. Such systems are complex in both their composition – typically many different kinds of components interacting simultaneously and nonlinearly with each other and their environments on multiple levels – and in the rich diversity of behavior of which they are capable.

The Springer Series in Understanding Complex Systems series (UCS) promotes new strategies and paradigms for understanding and realizing applications of complex systems research in a wide variety of fields and endeavors. UCS is explicitly transdisciplinary. It has three main goals: First, to elaborate the concepts, methods and tools of complex systems at all levels of description and in all scientific fields, especially newly emerging areas within the life, social, behavioral, economic, neuro- and cognitive sciences (and derivatives thereof); second, to encourage novel applications of these ideas in various fields of engineering and computation such as robotics, nano-technology and informatics; third, to provide a single forum within which commonalities and differences in the workings of complex systems may be discerned, hence leading to deeper insight and understanding.

UCS will publish monographs, lecture notes and selected edited contributions aimed at communicating new findings to a large multidisciplinary audience.

For further volumes:
http://www.springer.com/series/5394

Nigel Gilbert • Petra Ahrweiler • Andreas Pyka
Editors

Simulating Knowledge Dynamics in Innovation Networks

Springer

Editors
Nigel Gilbert
Department of Sociology
Faculty of Arts and Human Sciences
University of Surrey
Guildford, Surrey
United Kingdom

Petra Ahrweiler
EA European Academy of Technology and Innovation Assessment GmbH
Bad Neuenahr-Ahrweiler
Germany

Andreas Pyka
Lehrstuhl für Innovationsökonomik
Universität Hohenheim
Stuttgart, Germany

ISSN 1860-0832 ISSN 1860-0840 (electronic)
ISBN 978-3-662-51148-0 ISBN 978-3-662-43508-3 (eBook)
DOI 10.1007/978-3-662-43508-3
Springer Heidelberg New York Dordrecht London

Printed on acid-free paper

Springer is part of Springer Science+Business Media (www.springer.com)

Preface

In this preface, we provide a brief history of Simulating Knowledge Dynamics in Innovation Networks, best known as the "SKIN model". It all started with the joint conference of the Society for the Social Studies of Science (4S) and the European Association for the Study of Science and Technology (EASST) in 1996, which took place at the University of Bielefeld in Germany under the title "Signatures of Knowledge Societies" and was organised by the Institute of Science and Technology Studies (IWT). At this conference, Petra Ahrweiler, as a young postdoc, had organised two sessions on "Modelling Science" and had invited Nigel Gilbert, the best-known expert in this field, as keynote speaker. Several young scholars were speakers at these sessions, among them Koen Frenken and Paul Windrum. The meeting, where we met for the first time, created two main outputs: a Springer book containing most of the contributions as chapters (*Computer Simulations in Science and Technology Studies*, edited by Petra Ahrweiler and Nigel Gilbert 1998) and applications from some of the session participants for two projects in the European 4th Framework Programme, both of which were awarded. One of these projects was the SEIN project (an acronym for "Simulating Self-Organizing Innovation Networks"), which would generate the first version of the SKIN model. At the time of the application, Koen Frenken was a member of the Economics team of Paolo Saviotti at INRA/SERD in Grenoble and became our French partner. However, when the SEIN project started in 1988, he was already back at Utrecht and was replaced in Paolo's team by a German postdoc working in Grenoble, Andreas Pyka.

The SEIN project was undertaken by an interdisciplinary team of social science researchers from Science and Technology Studies, Sociology, Economics, and Political Science. The project combined five empirical case studies in different sectors of technological innovation and in different EU member states with agent-based simulation. The results of the SEIN project are summarised in Pyka and Kueppers (2003). The case studies described knowledge-intensive European industry sectors such as the biotechnology-based pharmaceutical industry in France (Pyka and Saviotti 2003), combined heat and power technology networks in the Netherlands, Germany, and the UK (Weber 2003), knowledge-intensive business services in the UK web design industry (Windrum 2003), and the UK Virtual

Centre of Excellence in the European telecommunication industry (Vaux and Gilbert 2003). The task of the SEIN project was threefold: theory formation, empirical case studies, and agent-based simulation. The objective was to derive a theory of innovation networks from insights derived inductively from the case studies and to implement this theory of innovation networks into an agent-based model.[1] The result of the modelling activities was an agent-based model, grounded in empirical research and informed by empirical data coming from the case studies (Gilbert et al. 2003). The model was used by the European Commission for scenario modelling of current and future innovation policy strategies (Ahrweiler et al. 2003) referring to the technological sectors and EU Member States of the case studies.

However, before this could happen, the project encountered a difficult situation: as often in European projects, towards the end of the project, time became more and more scarce and the promised agent-based model was not nearly finished. The consortium had outsourced the modelling activity to a US partner, the Swarm Corporation, an offshoot of the famous Santa Fe Institute in New Mexico. Here worked Glen Ropella, a wonderful mathematician with whom we had many nice project meetings. However, their "house model", the Swarm platform, while very appropriate for modelling simple reactive/adaptive agents, struggled with endowing agents with complex knowledge and heterogeneous properties while letting them follow complicated rules of behaviour and learning as outlined in our theoretical framework and observed in the case studies.

The agent-based model was supposed to be the running version of our theoretical framework, and the success of the project depended on it. Lacking faith that the Swarm model would be ever be able to do the job we had expected of it, Nigel Gilbert, Andreas Pyka, and Petra Ahrweiler met in the summer of the year 2000 at the University of Surrey in Guildford, UK, during the last half year of the SEIN project, to discuss what we could do to rescue the situation. In our favour was that we had learnt to listen and talk to each other in the course of the project although we were from different disciplinary backgrounds (sociology, science and technology studies, economics), we were all theory/model-oriented scientists, we already had relatively clear ideas on how to model our theoretical framework and empirical insights, and—last but not least—Nigel had experience in modelling plus a background in artificial intelligence programming in LISP. Against us was of course time, a certain frustration at having to do all the work ourselves without having been assigned to it in the first instance, having to spend the summer like this, and the restrictive EU travel reimbursement rules, which had the external visitors confined to the Guildford YMCA. We stayed for about 10 days and developed not only the model but many habits, which have stayed with us, such as going for long walks and hikes, visiting Haiku the cat, having morning coffees at Caffè Nero in Guildford

[1] Since empirical research is embedded in theoretical frameworks itself, this actually meant a close interaction of theoretical frameworks and empirical research from the outset of the project. The case study partners provided theoretical contributions from Neo-Schumpeterian Economics, Economic Sociology, Organisational Learning, etc. to guide empirical research in the case studies.

High Street to avoid the calorie-prone warm breakfast at the Y, and having regular morning runs along the local river Wye for the same reason. In 2001, the result of our efforts was published as the SEIN model (Gilbert et al. 2001). It already had the basic features of the later SKIN model.

SEIN had many procedures designed from the perspective of the biotechnology-based pharmaceutical industry in Europe, using this sector as an example *par excellence* of a knowledge-intensive industry and basing our knowledge of it on the empirical SEIN case study that Andreas Pyka had been responsible for. The first version of the model to include a market was developed during the bi-national research project of the British Research Council and the German DAAD "Comparing German and UK biotechnology-based pharmaceuticals: Simulating Knowledge Dynamics in Innovation Networks" (SKIN). The empirical work is summarised in Ahrweiler et al. (2006); the SKIN version with the market is first presented in Ahrweiler et al. (2004).

When we first sat in front of a blank whiteboard in 2000, we thought that our ideas about building an agent-based model might be used to understand the dynamics of innovation networks better, and perhaps create a larger community with which we can share our ideas and further develop this approach. As this book attests, our dreams about developing a model of innovation networks that could be the basis of a wide range of studies have come true. The scope of studies applying close relatives of our original 2000 model has already become very broad. We look forward with excitement to the next steps and to growing a community—even an innovation network—of those using agent-based models to shed light on the processes of innovation and with these modelling activities to enhance our respective disciplines.

Bad Neuenahr-Ahrweiler, Germany Petra Ahrweiler
Stuttgart, Germany Andreas Pyka
Guildford, UK Nigel Gilbert

References

Ahrweiler P, Gilbert N (eds) (1998) Computer simulations in science and technology studies. Springer, Berlin

Ahrweiler P, de Young S, Windrum P (2003) Evaluating innovation networks. In: Pyka A, Kueppers G (eds) Innovation networks: theory and practice. Edward Elgar, Cheltenham

Ahrweiler P, Pyka A, Gilbert N (2004) Simulating knowledge dynamics in innovation networks (SKIN). In: Leombruni R, Richiardi M (eds) Industry and labor dynamics: the agent-based computational economics approach. World Scientific Press, Singapore

Ahrweiler P, Gilbert N, Pyka A (2006) Institutions matter but … Organisational alignment in knowledge-based industries – the case of German and British bio-pharmaceuticals. Sci Technol Innovat Stud 2(1):3–18

Gilbert N, Pyka A, Ahrweiler P (2001) Innovation networks: a simulation approach. J Artif Soc Soc Simulat 4(3):8. http://jasss.soc.surrey.ac.uk/4/3/8.html

Gilbert N, Pyka A, Ahrweiler P (2003) Simulating innovation networks. In: Pyka A, Kueppers G (eds) Innovation networks: theory and practice. Edward Elgar, Cheltenham

Pyka A, Kueppers G (eds) (2003) Innovation networks: theory and practice. Edward Elgar, Cheltenham

Pyka A, Saviotti PP (2003) Innovation networks in the biotechnology-based industries. In: Pyka A, Kueppers G (eds) Innovation networks: theory and practice. Edward Elgar, Cheltenham, pp 75–107

Vaux J, Gilbert N (2003) Innovation networks by design: the case of mobile VCE. In: Pyka A, Kueppers G (eds) Innovation networks: theory and practice. Edward Elgar, Cheltenham

Weber M (2003) Innovation networks and the transformation of large socio-technical systems: the case of combined heat and power technology. In: Pyka A, Kueppers G (eds) Innovation networks: theory and practice. Edward Elgar, Cheltenham, pp 133–165

Windrum P (2003) The role of knowledge-intensive business services (KIBS) in e-commerce. In: Pyka A, Kueppers G (eds) Innovation networks: theory and practice. Edward Elgar, Cheltenham, pp 108–132

Contents

List of Contributors

Petra Ahrweiler EA European Academy of Technology and Innovation Assessment GmbH, Bad Neuenahr-Ahrweiler, Germany

Mercedes Bleda Manchester Institute of Innovation Research (MIoIR), Manchester Business School, University of Manchester, Manchester, UK

Martin Blom Center for Technology, Innovation and Culture, University of Oslo, Oslo, Norway

Tobias Buchmann University of Hohenheim, Stuttgart, Germany

Fulvio Castellacci Norwegian Institute of International Affairs, Oslo, Norway

Ozge Dilaver Centre for Research in Social Simulation (CRESS), University of Surrey, Guildford, UK

British Institute at Ankara, Ankara, Turkey

Arne Fevolden NIFU Nordic Institute for Studies in Innovation, Research and Education, Oslo, Norway

Nigel Gilbert Centre for Research in Social Simulation, University of Surrey, Guildford, UK

Jarle Moss Hildrum Center for Technology, Innovation and Culture, University of Oslo, Oslo, Norway

Manuela Korber Innovation Systems Department, AIT Austrian Institute of Technology GmbH, Vienna, Austria

Muhamed Kudic Department Structural Economics, Halle Institute for Economic Research, Halle (Saale), Germany

Christian Lüthje Institut für Innovationsmarketing (W-3), Technische Universität Hamburg-Harburg, Hamburg, Germany

Matthias Müller University of Hohenheim, Stuttgart, Germany

Manfred Paier Innovation Systems Department, AIT Austrian Institute of Technology GmbH, Vienna, Austria

Andreas Pyka Universität Hohenheim, Stuttgart, Germany

Michel Schilperoord UCD Innovation Research Unit (IRU), UCD Michael Smurfit Graduate Business School, University College Dublin, Carysfort Avenue, Blackrock, Dublin, Ireland

Benjamin Schrempf EA European Academy of Technology and Innovation Assessment GmbH, Bad Neuenahr-Ahrweiler, Germany

Stephanie Hintze Institut für Innovationsmarketing (W-3), Technische Universität Hamburg-Harburg, Hamburg, Germany

Elvira Uyarra Manchester Institute of Innovation Research (MIoIR), Manchester Business School, University of Manchester, Manchester, UK

Chapter 1
Simulating Knowledge Dynamics in Innovation Networks: An Introduction

Petra Ahrweiler, Andreas Pyka, and Nigel Gilbert

Abstract In this introduction, we outline the theoretical background for the most important concepts of the Simulating Knowledge Dynamics in Innovation Networks (SKIN) model. We describe the basic model, which we understand more as a theoretical framework than as a piece of code and preview the following chapters, which apply the SKIN model to diverse industrial sectors and develop related network models to generate insights about the dynamics of innovation networks.

1.1 Knowledge Representation in SKIN

Following the ideas of Neo-Schumpeterian Economics and Economic Sociology, we define innovation as the creation of new, technologically feasible, commercially realisable products, processes and organisational structures (Schumpeter 1911; Fagerberg et al. 2006). It results from the continuous interactions of innovative organisations such as universities, research institutes, firms such as multi-national corporations and small-to-medium-sized enterprises, government agencies, venture capitalists and others. These organisations exchange and generate knowledge by drawing on networks of relationships (innovation networks) that are embedded in institutional frameworks at the local, regional, national and international levels. The SKIN model represents these organisations as agents, embedded in a network.

P. Ahrweiler (✉)
EA European Academy of Technology and Innovation Assessment GmbH, Wilhelmstr. 56, 53474 Bad Neuenahr-Ahrweiler, Germany
e-mail: Petra.Ahrweiler@ea-aw.de

A. Pyka
Universität Hohenheim, 70593 Stuttgart, Germany
e-mail: a.pyka@uni-hohenheim.de

N. Gilbert
Centre for Research in Social Simulation, University of Surrey, Guildford GU2 7XH, UK
e-mail: n.gilbert@surrey.ac.uk

N. Gilbert et al. (eds.), *Simulating Knowledge Dynamics in Innovation Networks*, Understanding Complex Systems, DOI 10.1007/978-3-662-43508-3_1,

1.1.1 Kenes

The first property agents need to be endowed with is an individual knowledge base to model their most important resource: "Technological capital is the portfolio of scientific resources (research potential) or technical resources (procedures, aptitudes, routines and unique and coherent know-how, capable of reducing expenditure in labour or capital or increasing its yield) that can be deployed in the design and manufacture of products" (Bourdieu 2005: 194). To model this, we require a type of semi-structured knowledge space: a categorical system structuring the codified and explicit knowledge "portfolio" not only mentioned by Bourdieu, but also in many literatures using terms such as disciplines, domains, fields, capabilities, procedures, aptitudes, routines and so on. Additionally, we want a knowledge representation that could include know-how and expertise, that is, tacit and/or implicit knowledge features. Agents should be able to explore and navigate their knowledge space, to connect to other agents' knowledge spaces as far as learning and cooperation goes, and to construct new knowledge.

The approach to knowledge representation used in SKIN is similar to Toulmin's (1967) evolutionary model of knowledge production. This identified concepts, beliefs and interpretations as the "genes" of scientific/technological development evolving over time in processes of selection, variation and retention. Ackermann (1970) interpreted the works of Kuhn and Popper according to this perspective, allowing for different selection systems. In the SKIN model an analogical concept, the "kene" is used to represent the aggregate knowledge of an organisation. The kene concept was first introduced in Gilbert (1997) to model the dynamic structure of academic science and we adapted it to model the knowledge of firms. The individual knowledge base of a SKIN agent, its kene, contains a number of "units of knowledge". In the artificial space of a model, kenes consist of arbitrary bit sequences of indefinite length.

As an empirical illustration of how some of the knowledge of a firm can be mapped to a kene, the International Patent Classification (IPC) provides a hierarchical system of language-independent symbols for the classification of patents according to the different areas of technology to which they apply. IPC Codes allow the assignment of technological fields and competences mentioned in patents with so-called concordance tables (Schmoch et al. 2003) to identify industrial sectors. In this sense, the IPC Codes can be considered as coordinates of an empirical knowledge space and correspond approximately to the "units of knowledge" in the kenes.[1]

Each unit in a kene is represented as a triple consisting of a firm's capability, C, in a scientific, technological or business domain, its ability, A, to perform a certain application in this field, and the expertise level, E, the firm has achieved with

[1] The analogy is not exact, for the kene units can represent items of tacit knowledge (which can therefore not be made explicit as IPC codes) and can also represent non-technical knowledge, such as business strategies, marketing techniques and management competences, none of which are patentable.

$$\left\{\begin{matrix} C \\ A \\ E \end{matrix}\right\}, \left\{\begin{matrix} C \\ A \\ E \end{matrix}\right\}, \left\{\begin{matrix} C \\ A \\ E \end{matrix}\right\}, \left\{\begin{matrix} C \\ A \\ E \end{matrix}\right\}, \left\{\begin{matrix} C \\ A \\ E \end{matrix}\right\}, \ldots$$

Fig. 1.1 The kene of an agent

respect to this ability (see Fig. 1.1). For example, if representing patents, a kene's capabilities *C* correspond to the 3-digit IPC Codes that relate to broad technological fields, e.g. the code C07 stands for organic chemistry. One level below, on the 4-digit level, are the various occurrences within the capabilities domain, which correspond to the abilities *A*, e.g. the code C07K represents all technologies related to processes of preparation of peptides. As every capability *C* has many (in principle infinite) occurrences (abilities *A*), each 3-digit IPC Code has several specifications on the 4-digit level, and these can be augmented if new, so far non-classified technologies become available.

For almost every technological artefact or process, many capabilities, as well as different abilities within one capability domain, need to be applied (so-called combinatorial technologies, cf. Teece 1987). Therefore patents generally list several IPC codes. A firm's kene consists of a variable number of *C*/*A*/*E* units that together represent its artificial knowledge space. The units of knowledge in the kene are used in a combinatorial way to produce innovations.

1.1.2 Innovation

Another feature of the SKIN model is the procedure for doing innovation, i.e. generating 'innovation hypotheses'. Firms apply their knowledge to create innovation. "Most technology is specific, complex ... (and) cumulative in its development ... It is specific to firms where most technological activity is carried out, and it is specific to products and processes, since most of the expenditure is not on research, but on development and production engineering, after which knowledge is also accumulated through experience in production and use on what has come to be known as 'learning-by-doing' and 'learning-by-using'" (Pavitt 1987: 9). The special focus of a firm, its potential innovation, is called an innovation hypothesis. In the model, the innovation hypothesis (IH) is derived from a subset of the firm's kene units. The underlying idea for an innovation, modelled by the innovation hypothesis, is the source an agent uses for its attempts to make profits. By applying the knowledge in its innovation hypothesis, an agent increases its expertise in this area. This is the way that learning by doing/using is modelled. The expertise levels of the units in the innovation hypothesis are increased and the expertise levels of the other, unused units are decremented. Expertise in unused units in the kene is eventually lost and the units are then deleted from the kene; the corresponding abilities are "forgotten" or "dismissed" (Hedberg 1981).

1.1.3 Learning

The learning features of the model are grounded theoretically in the body of literature known as "Organizational Learning". After John Dewey introduced the concept of experiential learning as a permanent activity cycle (Dewey 1938) and started a discussion among educationalists about feedback learning and learning by doing, Donald Michael coined the term, organisational learning (Michael 1973). Argyris and Schön's influential monograph, "Organizational Learning" (1978; re-published with additional material as "Organizational Learning II", 1996) proposed that a learning organisation is one that is permanently changing its interpretation of the environment. In doing so, the organisation learns new things and forgets old ones. Drawing on their background as action theorists, Argyris and Schön show how these interpretations are conceived and how they are connected to different organisational behaviours. They distinguish between three types of learning, rooting them in an understanding of organisational agency that targets growth and effectiveness:

Single-Loop Learning This is adjustment learning, referring to the rational use of one's own means and instruments to adapt to environmental requirements, given a set of organisational goals, strategies and behaviours. It targets an improvement of the "theory in use" of an organisation using a simple action-outcome feedback and follows the heuristic, "maximise gains and minimise loss".

Double-Loop Learning This is turnover learning with respect to the meta-level of goals, strategies, and behaviours of an organisation, and aims to adapt these meta-level features to environmental requirements. The learning process includes un-learning of redundant knowledge to clear space for new behaviours. Furthermore, co-operation, including sharing assumptions and benefits with collaborators, is seen as a vehicle for learning.

Deutero Learning This is meta-level learning of the highest order where the organisation reflects on its own identity. Here, the learning process itself is the object of learning ("to learn how to learn"). The organisation's norms and values are subject to critique and change.

The SKIN model takes many of the ideas of the Argyris and Schön framework and uses them to examine the assumption that, in the words of de Geus (1997), the greatest competitive advantage for any firm is its ability to learn. Experiments concerning the effects of different combinations of learning activities on the agent population are reported in Gilbert et al. (2007). In the model, firms may engage in single- and double-loop learning activities. Firm agents can:

- Use their capabilities (learning by doing/using) and learn to estimate their success via feedback (learning by feedback) and/or
- Improve their own knowledge incrementally when the feedback is not satisfactory in order to adapt to changing technological and/or economic standards (adaptation learning, incremental learning).

If a firm's previous innovation has been unsuccessful, it considers that it is time for change. If the firm still has enough capital, it will carry out "incremental" research (e.g. R&D in the firm's laboratory). Performing incremental research (Cohen and Levinthal 1989) means that a firm tries to improve its product by altering one of the abilities chosen from the units in its innovation hypothesis, while sticking to its focal capabilities. The ability in each unit is considered to be a point in the respective capability's action space. To move in the action space means to go up or down by an increment, thus allowing for two possible "research directions".

Alternatively, firms can radically change their capabilities (innovative learning, radical learning). A firm agent under serious pressure and in danger of becoming bankrupt will turn to more radical measures, by exploring a completely different area of opportunities. In the model, an agent under financial pressure turns to a new innovation hypothesis after first "inventing" a new capability for its kene. This is done by randomly replacing a capability in the kene with a new one and then generating a new innovation hypothesis.

According to the Organizational Learning theoretical framework, firms may also be active on the double-loop learning level of the model. They can:

- Forget their capabilities (clean up their knowledge space).
- Decide on their individual learning strategies themselves (e.g. incremental or radical learning), constructing and changing the strategies according to their past experience and current context. The context consists of external factors such as the actions of clients, competitors and partners and the availability of technical options, and internal factors such as their capital stock and the competencies available to them, and
- Engage in networking and partnerships to absorb and exploit external knowledge sources, to imitate and emulate, and to use synergy effects (participative learning).

1.1.4 Networks

An agent in the model may consider partnerships (alliances, joint ventures, etc.) in order to exploit external knowledge sources. "Social capital is the totality of resources activated through a more or less extended, more or less mobilizable network of relations which procures a competitive advantage by providing higher returns of investment" (Bourdieu 2005: 194).

The decision whether and with whom to co-operate is based on the mutual observations of the firms, which estimate the chances and requirements coming from competitors, previous and potential future partners, and clients. Bolton et al. (2005), writing from a theoretical viewpoint, and Michelet (1992) using empirical evidence, both show that greater mutual information, where firms know their partner's history of co-operation, improves the conditions for co-operation.

In the model, a marketing feature provides the information that a firm can gather about other agents: to advertise its product, a firm publishes the capabilities used in its innovation hypothesis. Those capabilities not included in its innovation hypothesis and thus in its product are not visible externally and cannot be used to select the firm as a partner. The firm's 'advertisement' is then the basis for decisions by other firms to form or reject co-operative arrangements.

When experimenting with the model, one can choose between two different partner search strategies (Powell et al. 2005), both of which compare the firm's own capabilities as used in its innovation hypothesis and the possible partner's capabilities as seen in its advertisement. Applying the conservative strategy, a firm will be attracted to a partner that has similar capabilities; using a progressive strategy, the attraction is based on the difference between the capability sets.

Previously good experience with former contacts generally augurs well for renewing a partnership. If there is a firm sufficiently attractive according to the chosen search strategy (i.e. with attractiveness above the 'attractiveness threshold'), it will stop its search and offer a partnership. If the potential partner wishes to return the partnership offer, the partnership is established.

The model assumes that partners learn only about the knowledge being actively used by the other agent. Thus, to learn from a partner, a firm will add the units of the partner's innovation hypothesis to its own. For capabilities that are new to it, the expertise levels of the units taken from the partner are reduced in order to mirror the difficulty of integrating external knowledge as stated in empirical learning research (cf. Cohen and Levinthal 1989; Cantner and Pyka 1998). For partner's capabilities that are already known to it, if the partner has a higher expertise level, the firm will drop its own unit in favour of the partner's one; if the expertise level of a similar unit is lower, the firm will stick to its own version. Once the knowledge transfer has been completed, each firm continues to produce its own product, possibly with greater expertise as a result of acquiring skills from its partner.

If the firm's last innovation was successful, i.e. the value of its profit in the previous round was above a threshold, and the firm has some partners at hand, it can initiate the formation of a network. This can increase its profits because the network will try to create innovations as an autonomous agent in addition to those created by its members and will distribute any rewards back to its members who, in the meantime, can continue with their own attempts, thus providing a double chance for profits. Networks get the same amount of initial capital as other firms and can engage in all the activities available to other firms. The kene of a network is the union of the units from the innovation hypotheses of all its participants. If a network is successful it will distribute any earnings above the amount of the initial capital to its members; if it fails and becomes bankrupt, it will be dissolved.

1.1.5 Getting Rich and Going Bankrupt

"Financial capital is the direct or indirect mastery of financial resources, which are the main condition (together with time) for the accumulation and conservation of all other kinds of capital" (Bourdieu 2005: 194). The agents of the SKIN model have individual capital stocks. They can prosper and they can go bankrupt. When it is set up, each firm has a randomly assigned amount of starting capital (a few randomly chosen firms can be given extra capital). The capital is needed to do innovation and to improve the firm's knowledge base. It can be increased by successful innovation. If a firm uses all its capital without making any profits, it leaves the simulation.

If a sector is successful, new firms will be attracted into it, representing Schumpeterian competition by imitation. This is modelled by adding a new firm to the population when any existing firm makes a substantial profit. The new firm is a clone of the successful firm, but with its kene units restricted to those in the successful firm's advertisement. This models a new firm copying the characteristics of those seen to be successful in the market. As with all firms, the kene may also be restricted because the initial capital of a start-up is limited and may not be sufficient to support the copying of the whole of the successful firm's innovation hypothesis. The new firm's kene units begin with a low level of expertise, to reflect the fact that the firm is just starting to innovate.

1.1.6 The Market

The underlying idea for an innovation, modelled by the innovation hypothesis, is the source an agent uses for its attempts to make profits in the market. Because of the fundamental uncertainty of innovation (Knight 1921), there is no simple relationship between the innovation hypothesis and product development. To represent this uncertainty, we developed the following mechanism: the innovation hypothesis is transformed into a product through a mapping procedure where the capabilities of the innovation hypothesis are used to compute an index number that represents the product. The particular transformation procedure applied allows the same product to result from different kenes, which is not too far from reality where the production technologies of firms within a single industry can vary considerably (Winter 1984).

A firm's product, P, is generated from its innovation hypothesis as

$$P = \left(\sum_{IH} C_i A_i \right) \bmod N \tag{1.1}$$

where N is a large constant and represents the notional total number of possible different products that could be present in the market.

A product has a certain quality, which is computed from the innovation hypothesis in a similar way by multiplying the abilities and the expertise levels for each unit in the innovation hypothesis and normalising the result. Whereas the abilities used to design a product can be used as a proxy for its product characteristics, the expertise of the applied abilities is an indicator of the potential product quality. In order to realise the product, the agent needs some materials. These can either come from outside the sector ('raw materials') or from other firms, which generated them as their own products. Which materials are needed is again determined by the underlying innovation hypothesis: the kind of material required for an input is obtained by selecting subsets from the innovation hypotheses and applying the standard mapping function [Eq. (1.1)].

The inputs are each different and differ from the firm's own product. In order to be able to engage in production, all the inputs need to be available on the market, i.e. provided by other firms or be raw materials. If the inputs are not available, the firm is not able to produce and has to give up this attempt to innovate. If there is more than one supplier for a certain input, the agent will choose the one with the cheapest price and, if there are several similar offers, the one with the highest quality.

If the firm can go into production, it has to find a price for its product, taking into account the input prices it is paying and a possible profit margin. While the simulation starts with product prices set at random, as the simulation proceeds, a price adjustment mechanism following a standard mark-up pricing model increases the selling price if there is much demand, and reduces it (but no lower than the total cost of production) if there are no customers. Some products are considered to be destined for the 'end-user' and are sold to customers outside the sector: there is always a demand for such end-user products provided that they are offered at or below a fixed end-user price. A firm buys the requested inputs from its suppliers using its capital to do so, produces its output and puts it on the market for others to purchase. Using the price adjustment mechanism, agents are able to adapt their prices to demand and in doing so learn by feedback.

Thus, in trying to be successful on the market, firms are dependent on the market environment, which provides the required inputs, and on their innovation hypothesis, i.e. on their kene. If a product does not meet any demand, the firm has to adapt its knowledge in order to produce something else for which there are customers (Duncan 1974). Here, the learning and cooperation features are again connected. All in all, firms face a highly dynamic market environment, with both the availability and price of inputs, and the potential for sales, varying as the simulation proceeds and the other firms try to find market opportunities for their innovations.

1.2 A Tour Through the Book: Extending and Applying SKIN

The ten contributions to this book extend the SKIN framework to different industrial sectors or employ related network models in order to address new questions and generate new insights about the dynamics of innovation and learning networks.

Part I contains those contributions that focus on the firm level and analyse the complex interactions of innovation strategies. Part II moves beyond the firm level and addresses policy by demonstrating how agent-based models like SKIN can be employed to improve knowledge of policy interventions that are confronted with strong uncertainty. Part III deals with the broader cases of general purpose technologies, changes of technological regime and regional innovation systems. The contributions in Part III show how the SKIN model is able to improve understanding of the qualitative changes that go hand in hand with innovative transformations of an economy.

Chapter 2 by Martin Blom and Jarle Moss Hildrum entitled "Firm-level Business Strategies and the Evolution of Innovation Networks in the Nordic Internet Service Industry" explores how the firm-level adoption of specific innovation strategies affects the macro-level structure of innovation networks and knowledge bases within the Nordic Internet Service Provider sector. With the help of agent-based models and social network analysis, the evolution of industrial knowledge bases is visualized. The authors suggest that the adoption by many firms of an exploration-oriented innovation strategy facilitates a densely connected innovation network in which the participating firms build complex but increasingly similar knowledge bases. This seems to have positive implications for the economic performance of large and wealthy firms, but potentially negative effects for smaller specialized firms and possibly the industry's overall potential for generating novelty. By contrast, the adoption by many firms of a more conservative and exploitation-oriented strategy facilitates a less densely connected network structure in which the knowledge bases of the firms are less complex and more heterogeneous. This latter outcome appears to have more beneficial implications for smaller specialized firms, and for the potential of the industry as a whole to generate novelty.

Stephanie Toth and Christian Lüthje in Chapter 3, "The Evaluation of Value Chain Marketing Strategies: An Agent-Based Approach", address a promising strategy to increase the success of supplier innovations, namely Value Chain Marketing (VCM). When applying a VCM strategy, raw material suppliers enlarge their target group beyond direct customers and also include indirect customers. VCM processes can be organized cooperatively or non-cooperatively. An early integration of the manufacturer represents cooperative VCM. Raw material suppliers pursue non-cooperative VCM if the integration of the manufacturer only takes place at a later stage. Based on diverse case studies the authors show that the effectiveness of cooperative and non-cooperative VCM depends on the newness of supplier innovations and the overlap between the knowledge bases of the involved agents. The evaluation of the effectiveness as well as the efficiency of cooperative and non-cooperative VCM is undertaken with the help of an agent-based model. This model is intended as decision support for raw material suppliers and facilitates the overall evaluation of VCM strategies. The heterogeneous agents interact within the value chain. Each type of agent (raw material supplier, manufacturer and downstream customer) is specialized and starts the simulation with specific attribute values and with different knowledge bases conceptualized as kenes.

The evolution of inter-firm innovation networks driven by heterogeneous actors that follow a selection of well-defined cooperation strategies is analysed in the contribution by Matthias Müller, Tobias Buchmann and Muhamed Kudic in Chapter 4, "Micro Strategies and Macro Patterns in the Evolution of Innovation Networks: An Agent-Based Simulation Approach". They simulate the interplay of micro-level firm behaviour that shapes the macro-level of network patterns and the repercussions of macro-level evolution on micro-behaviour. The understanding of large-scale network patterns is important because the network topology affects the exchange of ideas and knowledge in networks. For instance, researchers have demonstrated that networks with short path lengths and high clustering—so-called small worlds—have a significant impact on the level of innovation. Thus, this study contributes to a deeper understanding of the systemic determinants of firm innovativeness by explaining the micro level drivers of macro level properties. Causal relationships between firm strategies and the emerging network structures are tested with the help of an agent-based simulation.

Part II of the book on testing policy options with agent-based network models is opened by Chapter 5 by Manuela Korber and Manfred Paier, "Simulating the Effects of Public Funding on Research in Life Sciences: Direct Research Funds Versus Tax Incentives". Building on the SKIN model, an agent-based model is specified that serves as a computational laboratory to address this issue. In the model, top-down funding is granted to all agent types if their research fields are in line with the specific topics defined by government. Indirect funding brings about tax benefits only for industry agents, no matter which research fields they hold. However, tax benefits affect small and medium sized enterprises and large enterprises in different ways, because the specific level of research expenses depends on an organization's size. Both theoretical underpinning and a solid empirical validation ensure applicability in the specific sectorial and spatial context. The credibility of the model is strengthened by a transparent calibration and validation strategy based on the authors' own qualitative and quantitative empirical research, as well as secondary data from Vienna, Austria. The simulation results anticipate crucial discussions in Austrian research policy and allow the *ex ante* assessment of the impact of future decisions about public research funding schemes.

Chapter 6 by Fulvio Castellacci, Arne Fevolden and Martin Blom is entitled "R&D Policy Support and Industry Concentration: A SKIN Model Analysis of the European Defence Industry". This chapter presents an agent-based model of the defence industry. The model simulates some of the key characteristics of the European defence sector and studies how firms in this market will respond to the challenges and opportunities provided by a higher degree of openness and liberalization in the future. The analysis suggests that European defence firms will progressively become more efficient, less dependent on public procurement and innovation policy support, and more prone to knowledge sharing and inter-firm collaborations. These firm-level dynamics will in the long-run lead to an increase in the industry's export propensity and a less concentrated market.

Another European topic is addressed in Chapter 7 by Petra Ahrweiler, Andreas Pyka, Michel Schilperoord and Nigel Gilbert, "Testing Policy Options for Horizon

2020 with SKIN". The authors describe how the SKIN model has been specialised to simulate the development of an intentional network: the European Commission's Framework 6 and 7 ICT theme, in which the nodes are funded projects and the links are created by the project partners. It is shown that with suitable adaptations and parameterisation, the model can fit the recent history of the ICT programme and can be used to assess the likely effect of a range of policy scenarios. This version of the SKIN model, INFSO-SKIN, includes agents that represent research institutions, large diversified firms and small and medium enterprises. Each of these has its own knowledge base that changes as a result of its activities in the network. The model simulates the research process from responding to a Commission Call by gathering partners and submitting a bid, though to carrying out research, producing deliverables and enhancing the collaborators' knowledge bases. A range of policy options are considered, for instance, increasing the size of consortia or insisting on the participation of small and medium enterprises, and the possible outcomes for the European Research Area are assessed using the model.

Part II of the book is completed with Chapter 8 by Michel Schilperoord and Petra Ahrweiler who extend the discussion on virtual policy laboratories. "Towards a prototype policy laboratory for simulating innovation networks" shows that major improvements are possible to the process of bringing together evidence from empirical studies and the results of simulation studies, the necessity of which has been addressed in literature on validation methods. This chapter develops a view on recent developments in designing and building computational policy laboratories based on the SKIN model that demonstrates that significant improvements in workflow are possible. The aim is to articulate the design of new components (tools) and workflow changes for a SKIN-based innovation policy lab. The chapter develops the key ideas by using examples from research on the Irish innovation system.

In Part III, the SKIN model is applied to broader questions addressing innovation and network dynamics in transformation processes of industries and whole innovation systems. Chapter 9, "Modelling the emergence of a general purpose technology from a knowledge based perspective—the case of nanotechnology", by Benjamin Schrempf and Petra Ahrweiler, addresses the nanotechnology sector as an example of general purpose technologies. With a broad range of applicability to many industries, the innovation networks for general purpose technologies differ considerably from those of other emerging technologies. The Irish case, with strong semi-conductor, materials, and biotechnology industries, characterized by the highest revealed technological advantages in Western Europe and a high share of nanotechnology patents, is a valuable example for investigating the dynamics of nanotechnology knowledge, its role in the economy, and the effect of policies on both. The chapter addresses the specific characteristics of general purpose technology innovation networks and suggest ways of modelling them.

The final chapter, Chapter 10, "Multilevel Analysis of Industrial Clusters: Actors, Intentions and Randomness (AIR) Model" by Ozge Dilaver, Elvira Uyarra and Mercedes Bleda, takes up the important regional dimension in innovation processes as reflected in the underlying innovation network structures. Interactions between agents have temporal as well as spatial properties. Regarding the former, interactions evolve into relationships with shared understandings, roles and expectations. Successful collaborations gradually form distinctive innovation networks, consisting of inter-personal and inter-organisational relationships and shared habits of thought that inform the nature of these relationships. Regarding the latter, it has been argued, particularly in evolutionary economic geography, that proximity between agents facilitates certain types of knowledge creation. Innovation networks, therefore, are often more closely-knit in some regions than in others. Furthermore, structural and evolutionary differences between regions may have important impacts on innovation networks. Overall, even in the absence of clear natural advantages, some regions emerge as centres of innovation, changing the socio-economic conditions and the development potential of the region significantly. Understanding the conditions for and the process of this emergence is an important research and policy objective. This chapter introduces several extensions to the SKIN model. Firstly, a multi-level modelling approach is adopted, incorporating not only firms as units of analysis but also individual actors and their interactions within the firm. This additional level of agency enriches the analysis by allowing the representation of innovative processes within firms. A multi-level understanding of these innovative processes is useful because the knowledge created during these processes exists in a distributed manner in the firm's codes, documents and procedures as well as in the minds of its employees. In addition, the approach enables representing, albeit simplistically, the organisational structure of the firm. The aim of the authors is to get a grasp of these multifaceted connections, bridging different research streams in organisational learning, evolutionary economic geography and computational social science. The elements of the agents' environment are represented with three spaces in the model, a reward space, a knowledge space and an earth space. In the reward space, the inherently uncertain nature of the returns to innovation and subjective understandings of these returns is tackled. In the knowledge space, types and properties of knowledge are addressed and the authors explore the creation of knowledge within and between firms. In the earth space, initial conditions, structures and institutions in a region are represented and the model allows one to follow the evolution of these regional elements.

References

Ackermann R (1970) The philosophy of science. Pegasus, New York

Argyris C, Schoen DA (1978) Organisational learning: a theory of action perspective. Addison-Wesley, Reading (re-published, 1996)

Bolton GE, Katoka E, Ockenfels A (2005) Cooperation among strangers with limited information about reputation. J Public Econ 89:1457–1468

Bourdieu P (2005) The social structures of the economy. Polity Press, Cambridge

Cantner U, Pyka A (1998) Absorbing technological spillovers. Simulations in an evolutionary framework. Ind Corp Change 7(2):369–397

Cohen WM, Levinthal D (1989) Innovation and learning: the two faces of R&D. Econ J 99: 569–596
de Geus A (1997) The living company. Brealy, London
Dewey J (1938) Experience and education. Collier, New York
Duncan RB (1974) Modifications in decision structure in adapting to the environment: some implications for organizational learning. Decis Sci 5:705–725
Fagerberg J, Mowery D, Nelson RR (eds) (2006) The Oxford handbook of innovation. Oxford University Press, Oxford
Gilbert GN (1997) A simulation of the structure of academic science. Sociol Res Online 2(2):3, http://www.socresonline.org.uk/2/2/3.html
Gilbert N, Ahrweiler P, Pyka A (2007) Learning in innovation networks: some simulation experiments. Phys A 378(1):100–109
Hedberg B (1981) How organizations learn and unlearn. In: Nystrom PC, Starbuck WH (eds) Handbook of organizational design. Oxford University Press, Oxford
Knight FH (1921) Risk, uncertainty and profit. Duncan, Chicago
Michael DM (1973) On learning to plan and planning to learn. Jossey-Bass, Hoboken
Michelet R (1992) Forming successful strategic marketing alliances in Europe. J Eur Bus 4:11–15
Pavitt K (1987) The objectives of technology policy. Sci Public Policy 14:182–188
Powell WW, White DR, Koput KW, Owen-Smith J (2005) Network dynamics and field evolution: the growth of inter-organizational collaboration in the life sciences. Am J Soc 110:1132–1205
Schmoch U, Laville F, Patel P, Frietsch P (2003) Linking technological areas to industrial sectors. Final report to the European commission. DG Research, Karlsruhe, Germany
Schumpeter JA (1911) Theorie der wirtschaftlichen Entwicklung, 8. Auflage, 1993. Duncker & Humblot, Berlin
Teece D (1987) Capturing value from a technological innovation. In: Guile BR, Brooks H (eds) Technology and global industry. National Academies Press, Washington, DC, pp 65–95
Toulmin S (1967) The philosophy of science: an introduction. Hutchinson, London
Winter SG (1984) Schumpeterian competition in alternative technological regimes. J Econ Behav Organ 5:237–261

Part I
Innovation Strategies

Chapter 2
Firm-Level Business Strategies and the Evolution of Innovation Networks in the Nordic Internet Service Industry

Martin Blom and Jarle Moss Hildrum

Abstract This chapter addresses how micro-level preferences and decisions about collaboration made by a large number of heterogeneous firms can affect the macro-level evolution of inter-firm networks and the structure of industrial knowledge bases. We conduct the analysis by way of an agent-based model (ABM) that mimics key stylized facts about firm-level business strategies and innovation networking in the Nordic internet service provider (ISP) industry.

To develop the model, we make use of an existing ABM—Simulating Knowledge Dynamics in Innovation Networks (SKIN) and to visualize the results we use the social network analysis software Gephi. As a means of improving the empirical validity of the model, we iteratively presented and modified the model through a sequence of discussions with strategic analysts at Telenor, a large Norwegian internet service provider.

The paper suggests that the adoption by many firms of a collaboration-oriented and explorative innovation strategy facilitates the evolution of a densely connected regional network in which the participant firms gradually build complex but increasingly similar knowledge bases. By contrast, the adoption by many firms of more internally oriented and exploitative innovation strategies facilitates the emergence of less densely connected networks in which the knowledge bases of the firms are less complex and more heterogeneous. This latter outcome appears to have more beneficial implications for the competitive performance of smaller specialized firms, and for the novelty-generation potential of the industry as a whole.

M. Blom (✉) • J.M. Hildrum
Center for Technology, Innovation and Culture, University of Oslo, P.O. Box 1108, Blindern, 0317 Oslo, Norway
e-mail: martin.blom@gmail.com; j.m.hildrum@tik.uio.no

N. Gilbert et al. (eds.), *Simulating Knowledge Dynamics in Innovation Networks*, Understanding Complex Systems, DOI 10.1007/978-3-662-43508-3_2,

2.1 Introduction

The objective of this chapter is to analyze how decisions about collaboration made by a large number of heterogeneous firms can affect the macro-level evolution of inter-firm innovation networks and the structure of industrial knowledge bases. We direct particular attention to visualizing the macro-level structure and dynamics of industrial knowledge bases.

Drawing inspiration from Grandori and Soda (1995, pp. 184–185), we refer to a regional industrial innovation network as a mode of regulating interdependence between regionally co-located firms that belong to the same industry. This mode of regulation is different from coordination through market signals and serves to facilitate the implementation of new products, services and processes. The definition encompasses a broad range of inter-firm collaborative relationships, such as the trade of technologies and formal R&D collaboration and the boundaries of the network are defined by geographical limits and industry characteristics. Throughout the paper, we refer to the existence of a collaborative relationship between two firms as a network tie and the individual firms involved as network nodes. We further characterize single network ties amongst two firms along a dimension of tie strength, which is a function of the number of such collaborative relationships as well as the duration and rate of renewal of each relationship.

A widespread argument in economics and management literature is that innovation networks constitute a critical source of competitiveness in firms (Teece 1992; Hagedoorn and Schakenraad 1994; Chesbrough 2003; Chesbrough et al. 2006; Dahlander and Gann 2010; McEvily and Zaheer 1999). The reason is that network membership can equip firms with valuable resources, like timely access to critical information, complementary knowledge and assets required to successfully take new products and services to market (Powell et al. 1996). On a more aggregate level, such networks can bolster the innovative capacity and competitiveness of regionally embedded industries. The topic of regionally embedded innovation networks is important for our understanding of firm performance, but it is also complex and involves multiple levels of analysis that are rarely analyzed in combination.

While some studies direct primary attention to the level of individual firms' strategic choices and how these affect the evolution of different network structures, other studies highlight the relationship between network structure and the economic performance of the participant firms.

Seeking to carry out integrated analyses that take multiple levels of analysis into account, several recent studies have turned to agent-based models (ABMs) as a primary method (Ozman 2007; Malerba and Orsenigo 1996; Breschi et al. 2000; Zirulia 2009; Antonelli 2011). These studies are united in that they explore how the micro-level diversity of firms in terms of business strategies affects the resilience of regional networks and firms' innovative performance. For instance, Antonelli and Ferraris (2011) present a model that examines how systemic knowledge interactions among a large number of heterogeneous firms in a network affects the chances of individual firms to introduce innovations. According to these authors, specific organizational features of a regional economic system, such as intellectual property particular regimes and infrastructure support, are likely to affect the long-term rate of technological innovation in the relevant region. Similarly, Lozano and Arenas (2007) use an agent-based model to explore how the diversity of firms in terms of knowledge and organizational characteristics affects the resilience of regional innovation networks.

While these previous ABMs have increased our knowledge about the evolving structure of ties between firms populating specific industries, less attention has been directed to the evolving structure of ties between different strands of knowledge possessed by the firms. As Rodan and Galunic (2004) have argued, we know too little about the structure of the knowledge content of inter-firm networks. The purpose of the present paper is to carry out an integrated analysis of firm level strategy, inter-firm networking and the evolution of industrial knowledge bases.

We conduct the analysis by developing an agent-based model (ABM) that mimics key stylized facts about one regionally embedded knowledge-intensive industry—the Nordic Internet Service Provider industry. This is a knowledge-intensive and fast-changing industry composed of firms that provide internet access and services in the Nordic market. The industry is characterized by a high level of innovation-oriented collaboration and thus serves our purposes as a basis for modeling. We believe that analyzing the structure and dynamics of industrial knowledge bases by way of an ABM is useful since it can give us new insight into how firm-level decisions about collaboration can affect the macro-level knowledge resources of a regional industry. We pose the following research questions:

1. How does the firm-level adoption of different innovation strategies affect the structural evolution of regionally embedded innovation networks?
2. How does the evolution of different kinds of inter-firm network structures affect the interlinking of heterogeneous knowledge resources possessed by the firms involved?
3. How does the evolution of different kinds of industry knowledge structures affect the economic performance of different categories of firms populating the networks?

To answer these questions, we make use of an agent-based model (ABM) that mimics some stylized facts of current innovation strategies of three different categories of ISP firms. We explore how these three distinct groups of firms interact, form networks and recombine knowledge in two scenarios that are similar in terms of key stylized facts but differ with respect to:

(i) The extent to which the firms carry out internal R&D activities
(ii) The firms' criteria (thresholds) for choosing to form a network tie with another firm
(iii) The specific criteria they use when selecting among potential partners with which to form a network tie

We direct specific attention to outcome variables of (a) the composition of links between network nodes, (b) individual firms' economic performance and survival, (c) the complexity and heterogeneity of the entire industry's knowledge base. Drawing on past research about innovation networks, we discuss the innovation potential of the industry in terms of the structure of network linkages and the heterogeneity of knowledge among the nodes. We have organized the remainder of the paper as follows: First, we briefly present past research on innovation networks and present the main conceptual building block of our analysis. Thereafter, we present our research methods and the empirical context upon which we build our model. The third and fourth section describes the model and presents the results. The final section offers a brief discussion and a conclusion.

2.2 Innovation Strategy, Network Structure, Knowledge Heterogeneity and Firm Performance

When developing innovations, firms often source knowledge resources available in local inter-organizational networks. A key argument in research on innovation networks is that the degree to which a network can facilitate innovation depends on the heterogeneity of knowledge possessed by the participant firms and the density of links between these (Uzzi 1997; Hansen 1999; Ahuja 2000; Burt 2004; Powell et al. 2005).

For instance, according to Schilling and Phelps (2007), dense local clustering that emerges through recurring and redundant tie-formation within a group of firms, can improve information transmission capacity in the network by efficient communication and trustful cooperation. Such communication is necessary in complex inter-firm innovation ventures that involve considerable exchange of tacit knowledge. Following a similar line of reasoning, Hansen (1999) demonstrates that dense and recurrent contact between R&D teams is conducive of complex knowledge integration. At the same time, local clustering can result in an unfavorable lock-in to certain ways of approaching problems, as it restricts in-flow of knowledge from the outside. As March (1991) has pointed out, long-term repeated interactions between a limited number of nodes can lead to increasing conformity and homogeneity of knowledge in the network and thereby fewer novel solutions.

By contrast, a more dispersed structure of weaker ties attributes the network with greater reach and greater potential for novelty creation by tapping a wider and more diverse range of knowledge resources (Uzzi 1997; Ahuja 2000). In this case, a number of firms can operate as bridges covering structural holes and connecting otherwise disconnected clusters of firms (Ahuja 2000). Such weak ties can be advantageous in terms of getting access to novel knowledge from diverse sources and facilitate innovation through exploration of new combinations (Hansen 1999; Ozman 2007). At the same time, high heterogeneity of knowledge can complicate the execution of joint innovation ventures since the involved parties are likely to operate with different technical languages, work procedures and code of conduct, and therefore experience difficulties in communication and collaboration (Noteboom 1999, 2002). According to Schilling and Phelps (2007), an efficient innovation network should encompass both dense local clusters of strongly tied firms and more distributed structures of weakly tied firms. Through this combination, the network obtains both the capacity to search for new knowledge outside the local context and to recombine, as well as to efficiently convert, that knowledge into innovations.

2.2.1 Firm Level Collaboration Strategies and the Evolution of Network Structure

The evolution of network structures is contingent upon the participating firms' strategies regarding R&D and external collaboration. An innovation network comes into being because a certain number of firms decide to undertake R&D and to collaborate with one another. In the following we develop a simple two-dimensional typology that describes some firm level strategies that are relevant in this context (see Fig. 2.1).

Drawing on Teece (1986, 1992), we refer to the first dimension as "collaboration orientation". This refers to firms' willingness to engage in development-oriented collaboration with other firms. On the one end of the dimension are firms that choose to undertake their innovation activities internally with no external collaboration. On the other end are firms that base their development activities on extensive collaboration with multiple partners. Building on James March (1991) ideas about exploration and exploitation, we call the second dimension *innovation orientation.* On one end of the dimension are firms that focus on incrementally improving their existing knowledge, with no ambitions to venture into new territory. On the other end are firms that seek to explore new knowledge and change their existing products and services.

Firms matching quadrant 1 seek to improve their existing competencies and typically carry out their development activities in-house. When they do collaborate, they choose partners with knowledge that is similar to their own. If many firms in a regional industry choose this strategic approach, the result will be a disintegrated network with low density of linkages and little new knowledge creation.

Firms that match the characteristics of quadrant 2 seek to develop entirely new knowledge and do so mainly in-house. When they do collaborate, they typically select partners with knowledge that differ from their existing knowledge. If many firms choose this strategy, the result will be a low-density network with many isolated pockets in which entirely new knowledge is created.

Firms matching quadrant three seek to refine their existing competencies through extensive collaboration with multiple partners. When selecting external collaboration partners, however, these firms tend to select firms with knowledge

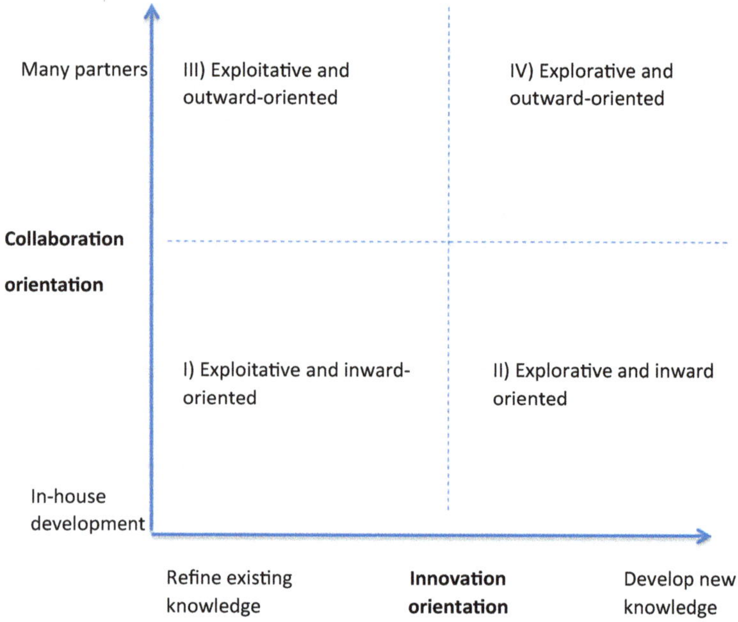

Fig. 2.1 A typology of four innovation collaboration strategies

that is similar to their own. If many firms in a regional industry choose this strategy, the result will be a densely connected network with little novelty and few new combinations of knowledge.

The last quadrant four describes firms that seek to develop entirely new knowledge through extensive collaboration with multiple external partners. When selecting among potential partners these firms tend to choose firms with knowledge that is different from what they already possess. If many firms adopt this strategy, the result will be a densely connected network with many new knowledge combinations. Over time, however, the network knowledge base might become increasingly homogeneous. Since there are many communication channels between the nodes, a normalization process might set in and few isolated pockets of knowledge remains.

In the remainder of the paper, we build an ABM model to explore how the adoption of two different strategies affects the build-up of industrial inter-firm networks, industrial knowledge bases and firm-performance.

2.3 Research Methods

As we pointed out above, this paper addresses how micro-level decisions about collaboration made by a large number of heterogeneous firms can affect the overall evolution of an inter-firm network. A useful way of exploring this topic is by way of agent-based models, in which a key function is to reveal dynamic relations between micro-level individual motives and macro-level social behavior (Taber and Timpone 1996; Axelrod 1997; Gilbert 2008). Compared to other methods the main benefit of using ABMs is that it allows us to undertake a dynamic and multilevel analysis of innovation networks.

To build our own model, we took as a starting point an existing agent-based model—the Simulating Knowledge Dynamics in Innovation Networks (SKIN) model which has been extensively used to simulate innovation networks in other industries, such as biotechnology and e-commerce (Gilbert et al. 2001, 2007, 2010; Pyka et al. 2007; Ahrweiler et al. 2011). An important benefit of this model, which is based on evolutionary economic theory about innovation, is that it allows agents to gradually learn and update their innovation collaboration strategies over time.

To modify the SKIN baseline model in such a way that it could be helpful in answering our research questions, we started by gathering stylized facts about the Nordic ISP industry. We consulted four main sources; (1) research literature and white papers about the Nordic ISP industry (Edquist 2003; Jansson 2011), (2) the European Internet Registry's (RIPE)[1] online database of Internet service providers, (3) online business registries made available by Nordic national statistical offices,[2] (4) an analysis of Norwegian ISPs responses to the EU's biannual community innovation survey (CIS) between 2004 and 2010, and (5) in-depth case studies of four Nordic ISPs. After we had developed a first baseline of the model, we calibrated it further through six meetings with four strategic analysts at Telenor, which is the largest Norwegian ISP. During the meetings which took place between December 2011 and March 2012 and lasted for 2 h each, we presented different versions of the model and asked for feedback on the extent to which the model reflected actual business strategies of Nordic ISPs. The results from these discussions constitutes an important contribution of the paper in the sense that it gave us insight into the nature of the Nordic ISP industry, as it is experienced by industry insiders.

We used the social network analysis tool Gephi to visualize network linkages and the heterogeneity of knowledge in the network. We will describe this model more in detail in the following sections, but first it is useful to describe the empirical setting which forms the starting point of the analysis.

[1] The Réseaux IP Européens Network Coordination Centre (RIPE) is the Regional Internet Registry (RIR) for Europe. As a RIR, RIPE oversees the allocation and registration of Autonomous System (AS) Numbers) in European countries. This is relevant in the context of our study since firms need AS numbers in order to independently sell Internet access capacity, and thus operate as ISPs. On RIPEs website (http://www.ripe.net/), there is a publicly available database of organizations with AS numbers that operate in the Nordic countries.

[2] These are Statistics Norway, Statistics Finland, Statistics Denmark and Statistics Sweden.

2.4 The Nordic Internet Service Provider Industry: Delineation and Stylized Facts

According to Lave and March (1993) the starting point of any good modeling project is some outcome in the empirical world that piques one's curiosity. The purpose of modeling then is to identify a process set in motion by specific micro-level preferences and behavior that plausibly explains the observed outcome. The empirical outcome that piqued our curiosity is a peculiar constellation inter-firm collaboration that currently characterizes the Nordic ISP industry, which has been described by Jansson (2011). The industry comprises a group of large established firms that conduct relatively little external collaboration, and a fast growing group of younger smaller firms with extensive inter-firm collaboration activities. We are interested in knowing more about how firm-level business strategies can affect the formation of inter-firm networks and knowledge bases in this industry, but we do not attempt to accurately replicate an observed empirical pattern through our analysis. Rather, we build our model on broad stylized facts about the industry environment and business strategies that firms in this industry have adopted in the past. The purpose of this exercise is to observe what happens on the network and industry knowledge base macro-level when we make slight alterations to these business strategies. The results, we hope, can be useful as a background for discussing past events as well as future scenarios in the Nordic ISP industry.

2.4.1 Delineation of the Industry

Following the European Network and Information Security Agency (ENISA) (2011, p. 82), we define an Internet Service Provider as a firm that offers direct Internet access and associated services to businesses and individual consumers. By associated services, we are referring broadly to consumer and business services which are offered on the Internet, such as online hosting of data and business software and online media content provision. The Nordic Internet Service (ISP) industry, then, is the population of Nordic firms that offer these services. As a further restriction, we counted as Nordic ISPs only those Nordic firms that possess their own autonomous system numbers (ASN) and thus have the ability to independently control Internet traffic on their own networks, and thereby sell and buy Internet access capacity with other firms or end-consumers.[3]

[3] An autonomous system is one Internet protocol (IP) network or sets of networks under a single administrative control. Companies and organizations might own more than one autonomous system, but the idea is that each autonomous system is managed independently. Autonomous System Numbers (ASNs) are globally unique numbers that are used to identify autonomous systems (ASes) and which enable an organization to route its own Internet traffic and to trade

To map the population of Nordic ISPs, we accessed an online database made available by the European Internet Registry (RIPE) which identifies all Nordic organizations that have been assigned an ASN (McGregor et al. 2010, see also http://www.ripe.net). It was possible to separate out organizations that operate in the Nordic market, due to a publicly available survey (also on the RIPE website) that shows the various geographical markets in which these organizations deliver their services and products. We cross-checked the resultant list against online business registries made available by national statistical offices of Norway, Sweden, Finland and Denmark to screen out non-Nordic firms and organizations that do not deliver Internet services as defined above. After this screening, we ended up with a list of 1,100 firms that currently provide Internet services of various kinds in the Nordic market. These firms constitute the target population of our analysis.

2.4.2 Stylized Facts

High innovativeness The Nordic ISP industry is characterized by a high level of innovativeness. From the mid-1990s until the turn of the millennium, the Nordic countries emerged as a world-leading region on development and use of Internet services, such as high-speed Internet access, wireless broadband, mobile multimedia services and online banking (Fraas et al. 2001; Hauknes and Smith 2002; Sjölander et al. 2001). One key aspect of this system is that there is a well-developed demand for advanced Internet services and the market is highly competitive for firms willing to respond to user needs (Edquist 2003; Jansson 2011). The Nordic countries also have some of the most liberal telecom & Internet markets in the world, which has been a major driver of the development of alternative networks (and Internet access provision) to the ones of the former incumbent national telecom carriers (Fjeldstad et al. 2004).

Dynamic structure and firm heterogeneity The industry is highly dynamic with high entry and exit rates, which has to do with rapidly emerging business opportunities, high investments and risks associated with the evolution of the Internet. During the last decade and a half, the Nordic ISP industry evolved from an early phase of domination by a handful of newly privatized national telecom carriers, through a period of strong diversification and entrepreneurial activity and the resultant emergence of a much more heterogeneous set of firms specializing in a wide set of technologies like Voice over IP, mobile videoconferencing, online hosting of data and business software, multimedia streaming, social media and integrated digital service platforms.

Internet access capacity. ASNs are administered and distributed by five non-profit organizations called regional Internet registries (one for each continent).

2.4.2.1 Stylized Agents

Access ISPs This is an ISP whose business is predominantly to sell Internet access to end users. The former national telecom carriers Telia-Sonera, Telenor and TDC, dominate this category. As a consequence of the liberalization of the Nordic telecom market, a large number of younger firms entered this end of the market during the last decade and a half, competing with the incumbents on Internet access quality and price. A key objective of access ISPs is to secure fast, reliable and high quality Internet access to their customers with as low costs as possible. This is primarily a matter of building high quality telecommunication networks and developing efficient methods for managing Internet traffic over these.

To achieve this objective, many Nordic access ISPs have adopted what we refer to as an inward-oriented and exploitative innovation strategy. By exploitative, we are referring to the fact that these firms rarely explore entirely new technologies or combinations of knowledge, but rather exploit technology and knowledge they already possess with a view to improving network operations (Cunha 2009). As Lechler et al. (2007) put it, 'while managers (of incumbent access ISPs) state that innovation is critical to the success of the carrier business, the firms' business objectives and metrics tell a different story.' Lechler's study shows that the firms, while commenting on the importance of innovation to stay competitive, really emphasize incremental improvement of existing network operations. By inward-oriented, we are referring to the fact that access ISPs typically carry out these innovation activities internally with relatively little actual R&D collaboration with external firms. When access ISPs need to collaborate with external partners they typically address long-standing suppliers of network equipment and software which are necessary to build and improve the networks (Edquist 2007). The knowledge bases of these external equipment suppliers are in most cases relatively similar to those of the access ISPs knowledge, which means that the collaborations are not likely to lead to entirely new combinations of knowledge.

Content ISPs A content ISP is an organization whose prime business is selling Internet access to external organizations that run websites, usually also selling a variety of data hosting and business services to external organizations as well. Some content ISPs combine these business services with end-user Internet access provision. As a consequence of this diversified source of income, the content ISPs typically carry out innovation activities across a wider range of technical areas than the access ISPs. Typical innovations are new ways of bundling together existing business services into integrated service packages, such as combinations of wireless Internet access, data security management and online hosting of data and business software. To develop these innovations, many content ISPs have adopted what we refer to as an outward-oriented and explorative innovation strategy. By explorative, we are pointing to the fact that many of these firms have experimented with entirely new combinations of services such as mobile videoconferencing systems, hosting and maintenance of business software, wireless broadband access and Internet content provision (Jansson 2011). By outward-oriented, we refer to the fact that

Table 2.1 Firm types in the Nordic ISP industry

Type	Approximate no of firms	Primary activities	Example
Access ISPs (out of which 5% are incumbents)	275	Internet access provision	Telenor, Telia-Sonera, TDC
Content ISPs	275	Internet content provision, web hosting, Internet access	Phonera, NetxGenTel
Specialized Internet service firm	550	Internet content production & provision, web-applications	Schibsted, TV2

that content ISPs frequently carry out these innovation activities through various kinds of collaborations with external firms. Indeed, as Jansson (2011) have pointed out, small and medium sized Nordic content ISPs typically search for new business opportunities through alliances with external firms with knowledge complementary to their own. As a consequence, many of the content ISPs these have a comparatively large portfolio of collaborations in quite different technical areas.

Specialized Internet service firm The specialized Internet service firm (SIS) focus on one or a few Internet services that are directly complementary to the ones offered by the access- and content ISPs. Examples are firms that offer Internet security services, provision of online content such as digitized books, music or movies. The specialized Internet service firms do not sell Internet access directly to end customers or businesses, but they frequently collaborate with access and content ISPs to market and deliver their services.

As there are no established industry classification standards for different Internet service providers, we had to make approximations regarding what proportion of firms in our sample fit into the above three categories. To do this, we searched the online public business registry proff.no and company websites to identify the main services of the firms. Since some firms do not fit neatly into either category, we also asked the strategic analysts at Telenor to indicate roughly how many firms should be allocated to the different categories. Our approximation is reported in Table 2.1.

While the dynamic character of the industry and use of collaborations are well known, we know relatively little about how firm level strategies and decisions affect the macro-level evolution of inter-firm networks and the capacity of the industry as a whole to generate novelty. Based on the above stylized facts, we develop an agent-based model to explore this relationship further.

We derive two scenarios from this model. In scenario 1 we keep all parameters in line with observed business strategies concerning cooperation between firms and degree of R&D. Access ISPs tend to be conservative in choice of cooperation partners and not so focused on developing brand new combinations of technologies and services. This is an *inward-oriented and exploitative business strategy*. Content ISPs on the other hand tend to be more open and progressive concerning cooperation and development of new services, representing an *outward-oriented and explorative business strategy*. The SIS firms have a position in between. With these strategies scenario 1 produces innovation network patterns for us to study. Scenario 2 represents a counterfactual simulation. What would happen if access ISPs were to adopt a more outward oriented and explorative innovation strategy? In this scenario all parameters are kept equal to the ones in scenario 1, except we change the access ISP's strategies. In scenario 2 access ISPs are almost as open and aggressive in their view on cooperation and service development as content ISPs. With a change in two parameters we study the change on macro level indicators for the industry and the innovation network patterns.

2.5 A Model of the Nordic ISP Industry

In order to describe the Nordic ISP industry with emphasis on innovation networks and explore the effects of firm-specific strategies, we employ an agent-based model (Gilbert 2008). In particular we start out with the Simulating Knowledge Dynamics in Innovation Networks (SKIN) model (Gilbert et al. 2001) and modify this to mimic the stylized facts of the Nordic ISP industry.

The SKIN model is an agent-based model that allows us to simulate firms' interactions based on innovation, learning and knowledge dynamics in knowledge intensive industries. In the following sections the structure and calibration of the model is presented, starting with Fig. 2.1 (below) which explains how one firm behaves in any given time period t.

2.5.1 Agents

The agents are responsible for the micro-level dynamics and emergent behavior on the macro-level. Each agent represents a firm in the Nordic ISP industry. In the model there are three types of agents, which makes it differ from the basic SKIN model: Access ISP's, content ISP's and SIS firms. These three types of firms are distinguished in the model according to size, service domain (available capabilities) and firm-level strategies like choice of partners and degree of service R&D. The access ISP's are firms which provide Internet access to end-users, but no hosting capabilities (Telenor) or other content and products. The content ISP's deliver hosting services, and in some cases also Internet access to end-users. The SIS firms deliver content services, but no Internet access and no hosting services.

The agents have in common that they all have a knowledge base represented by 'kenes'. Each kene is composed of three elements:

Capability (C): The capability represents a technological domain, or area of knowledge, within the Nordic ISP industry. In this model there are 300 different capabilities defined. This is a number that relates to the number of firms and is chosen to ensure a well-functioning market in the model. Too many capabilities in relation to the number of firms and it would be difficult for the firms to cooperate. The firms' knowledge bases would be too dissimilar.

Ability (A): The ability is more competence related and defines the ability to perform a certain application in this area of knowledge. To transform knowledge into a service, a way of employing this knowledge is needed. This could for instance be a type of facility (for data storage and hosting) or ISP infrastructure. Here it is represented as an integer between 1 and 10, where a change in the number represents a change in the facility (ability).

Expertise (E): The expertise is an indication of how well the firm masters this ability. When a kene is used repeatedly, the expertise level changes. This incorporates 'learning by doing' in the model. It is defined as an integer between 1 (low) and 10 (high).

A firm can have several capabilities, each with its own corresponding ability and expertise. These kenes are randomized at the beginning of the simulation in order to ensure randomness and heterogeneity in the population.

As explained above, the Nordic ISP industry is characterized by a high degree of technological change. The model assumes firms use and tries to improve their knowledge base in order to produce a better service which meets their profit target. The service is made on the basis of some of the kenes in the knowledge base. These make up what is called the Innovation Hypothesis (IH). The quality of the product is characterized by the abilities and expertise in the IH and this is a measure of the depth or specialization. Capabilities in the IH which are used several times, gain a higher expertise level. Capabilities in the knowledge base which are not included in the IH, loose expertise and are eventually forgotten.

2.5.2 *Market Environment*

Reflecting the stylized facts, the market environment in the simulation replicates stylized facts about the Nordic ISP market. The firms operate in a market where they provide services which work as input to other firm's services. In this market they compete on the basis of price and quality of service. A firm which cannot find the appropriate inputs to its own service to an affordable price will not be able to provide the service. The inputs a firm needs are dependent of the characteristics of the service and they find them by searching through other firms' services, prices and quality. In this market there are adjustment mechanisms for pricing the service. If there is much demand the price increases and vice versa.

The market is divided among the firm types. The content ISP's has access to all capabilities, whereas the access ISP's are excluded from the hosting capabilities. The SIS firms are excluded from both hosting and end-user Internet access capabilities but have access to capabilities related to general Internet activity and content services.

In line with the stylized fact *dynamic structure and firm heterogeneity*, the model introduces and removes a small percentage of firms from the population each time period.

2.5.3 Innovation and Partnerships

In the model, innovation is a driving force behind the economic dynamics and linked with the market performance of a firm. If a firm is not happy with its performance according to a given success-threshold, it has two options: (1) Innovation through internal service R&D. When defining service R&D we are aware ISPs do very little formal R&D as it is for instance defined in the OECD Oslo. However they do a lot of development in the sense that they bundle services in new ways, come up with different pricing strategies etc. (2) Innovation by forming partnerships with other firms (inter-firm collaboration), which include external learning and elements of imitation. These strategies are associated with uncertainty, since the result is not necessarily positive, and this introduces another random element in the model. Internal service R&D is done when there is demand for the service, but the profit is too low. It works by adjusting the ability in one of the kenes of the IH. Entering partnerships is chosen as a strategy when there is no demand for the service and the firm in question needs external knowledge from other successful firms. Cooperation works by adding a kene from the partner firm into its own IH, but with a reduced expertise level. Firms will choose to either do internal service R&D or forming partnerships as long as their service does not meet the profit target which is the parameter 'success-threshold'.

To summarize the model (see Fig. 2.2), two key properties describe the model. On the one hand, in any period t, firms operate within two different groups (loops). Loop 1 characterizes firms which are successful innovators and that sell their services. Loop 2 characterizes firms that are unsuccessful performers and that engage in either internal service R&D or innovation based on partnerships and in-sourcing of knowledge. Secondly, there are two different cumulative causation mechanisms. One where good performers increase their expertise and further strengthen their position in the market, and another catch-up mechanism where bad performers have the chance to improve their performance through innovation based on in-house R&D or partnering with other firms.

2.6 Variables and Calibration of the Model

2.6.1 Key Parameters

There are two key variables of particular interest in our analysis. They represent firm-level strategies and shape the market environment. We explore what happens

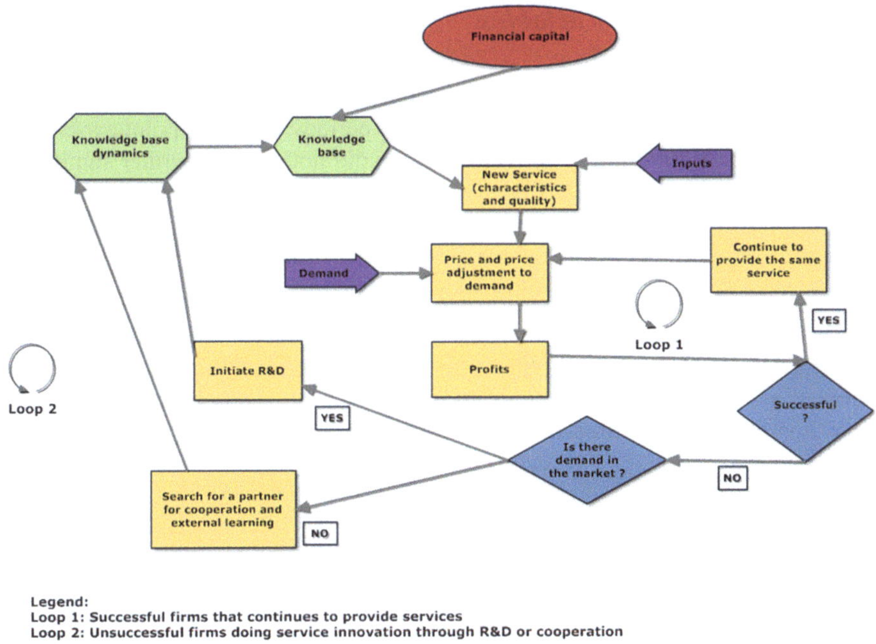

Fig. 2.2 A flowchart that describes the behavior of a firm in the SKIN model, in any given period t

to the whole market when a subset of firms in the same market changes their strategies, and thus changes the value of these parameters.

Cooperation This variable defines how similar the kenes of a possible partner must be in order for a firm to enter into a partnership. The value of the parameter ranges from 0 to 1. A value of 1 indicates that the potential partner must have a 100% identical knowledge base, whereas the value 0.5 indicates the potential partner must only share 50% of the knowledge base in order to enter into a partnership. Higher values represent an inward-oriented and exploitative strategy and tend to inhibit firms with differing kenes to cooperate and exchange knowledge.

Success threshold This variable defines the price a firm minimum aims to achieve for a service. If the revenue is higher than the threshold, the firm will continue to deliver and sell the service. Should the price be lower, the firm will seek to innovate through R&D by making adjustment to the kenes of their service. The parameter is defined in the profit range 0 to 1,000. The effect of a higher value is a higher degree of innovation and possibly a more open and competitive market.

2.6.2 Key Outcome Variables

The results presented for each variable is an average of 50 simulations done for each scenario. Each simulation was run for 200 time periods, at which point experience (several runs past 200 time periods) told us we did not see any further significant change in the trend already visible. The number of firms, concentration index, average innovation hypothesis (IH) length and the variance are all presented as a function of time. The subsequent variables are not time dependent, but an aggregated average for the 200 time steps, and still an average of 50 simulations.

Number of firms There are 1,100 firms at the start of each simulation. 275 access ISPs, 275 content ISPs and 550 SIS firms. The number of firms varies during the course of a simulation, due to the death of firms and the addition of new start-up firms. For each of the three firm-types the number of firms is reported at any given time.

Degree The degree reports the average number of partners a firm has at a given point in time. A higher (lower) degree means a more (less) connected network.

Weighted degree The weighted degree takes into account the number of times a firm has been in contact with a specific partner during a specified time period. A higher weighted degree means each partnership has been renewed more often during the course of the 200 time periods. The weighted degree reflects the strength of a tie between two firms, according to our earlier definition of tie strength as measured by the duration and intensity of contact between employees in the two firms in question.

Cluster coefficient The cluster coefficient is a measure of the degree to which firms (nodes) tend to cluster together, that is the extent to which subsets of firms tend to form ties with one another. The algorithm employed here is the network average clustering coefficient, meaning it is the average of all the local individual cluster coefficients. The local cluster coefficient is the number of edges[4] in a node's neighborhood divided by the possible number of edges in a nodes neighborhood, defined between 0 and 1.

Modularity coefficient (Q) The modularity coefficient Q is a measure of the degree to which a network is divided into sub-communities (clusters). It is defined as a sum, where the number of edges in a community is subtracted by the expected number of edges in a random such community, and this is aggregated for all identified communities. The value is defined between -0.5 and 1. Positive values signify a degree of community structure in the population. By community structure we mean that we have a subset of firms which tend to form ties with one another, just like a cluster. A pair of nodes (firms) is more likely to be connected with an

[4] An edge is a link between two nodes (firms). A tie, as defined earlier, between two firms is represented by an edge between two nodes.

edge (tie) if they belong to the same community. Using the modularity coefficient, we identify these communities (clusters) and are able to visualize the clusters separately from each other like in Figs. 2.4 and 2.5. If there is less community structure in a population, than in a random network of equal size and degree with randomly scattered edges, the modularity coefficient is negative. These results may be hampered by the resolution limit, though it is not certain (Fortunato and Barthelemy 2007). As part of the algorithm a number of modularity communities are identified. These are the perceived clusters of nodes (Blondel et al. 2008).

Number of triangles Three nodes with three edges connecting the nodes are considered a triangle. A node may participate in several triangles. This presents an indirect measure of clustering. Due to the resolution limit problem, the number of triangles is included to strengthen the reliability of the number of modularity communities.

Jaccard index This is a statistical value based on the Jaccard similarity coefficient. The Jaccard similarity coefficient is used to estimate the degree to which units in a sample are similar or different along a set of variables. It is defined between 0 and 1, with 1 being identical. A high value means that a specific set of values along a set of variables are typically observed together in units of observation (in our case firms). We use the 300 capabilities identified earlier as variables and services as observations. As a consequence we obtain a 300×300 matrix with Jaccard similarity coefficients. The reported Jaccard index is an aggregation of all the similarity coefficients (minus 300, the 300 identity values).

2.6.3 Model Calibration

There are a range of parameters which shape the model environment and make it resemble the Nordic ISP industry. The choices of value for these parameters are based on published research literature about the Nordic ISP industry, conversations with strategic analysts in Telenor, and on our own empirical data about the industry. In addition, there has been performed sensitivity analysis on the central parameters where necessary, helping to deduce an initial value.

Empirically based data:

The overall number of firms is based on the empirical data from the European Internet Registry RIPE (http://www.ripe.net/). As we pointed out above, RIPE has a database available on its website which enlists all firms which have applied for and been dealt an autonomous system number (ASN), and thus has the capacity to independently trade Internet access capacity. The number of firms belonging to each of the three categories (access ISP, content ISP and SIS firm) is derived from our approximations using business registries made available by national statistical offices in Norway, Denmark, Sweden and Finland.

The following parameters are calibrated based upon our conversations with strategic analysts in Telenor: The big firm percentage, cooperation parameter,

Table 2.2 A table of model parameter values

Access ISPs	
Number of firms	275
Cooperation	0.80
Success threshold	600
Initial capital	40,000
Big firm percent	5 %
Content ISPs	
Number of firms	275
Cooperation	0.35
Success threshold	1,000
Initial capital	20.000
Big firm percent	0 %
SIS firms	
Number of firms	550
Cooperation	0.40
Success threshold	800
Initial capital	20,000
Big firm percent	5 %
Global parameters	
Number of services	800
Number of inputs	4
In-out service percent	15 %
Reward to trigger start-up	20.000
Random start-up %	0.45 %
Random death %	0.1 %

success threshold, initial capital and random start-up- and death percentage. These parameters were calibrated by being given an initial estimate of their value by the strategic analysts at Telenor in our conversations with them. Then we ran the model and readjusted these values to reflect the stylized facts at macro-level.

The rest of the model parameters derive from a trial-and-error approach where we aimed at reproducing the stylized facts at the macro level.

Table 2.2 contains the parameter values of scenario 1. It reflects that the access ISPs have a low degree of service innovation and partnering activity, (or inward-oriented and exploitative business strategy), but that they are more capital intensive in general. These firms are larger, more incumbent firms. The content ISPs have a high degree of service innovation and partnering activity, or outward-oriented and explorative business strategy, but with less initial capital. They are up-and-coming small & medium sized firms. The SIS firms have a moderate level of service innovation and partnering activity.

In scenario 2 the setting is almost identical. However, the access ISPs adopts a more outward-oriented and explorative strategy towards cooperation (external imitation and learning) and internal R&D. To achieve this change in business strategy, the cooperation parameter changes from 0.80 to 0.40 (equal to the SIS firms' value, making it easier to cooperate) and the success threshold parameter from 600 to 900 (set to be higher than the value for the SIS firm, but lower than the

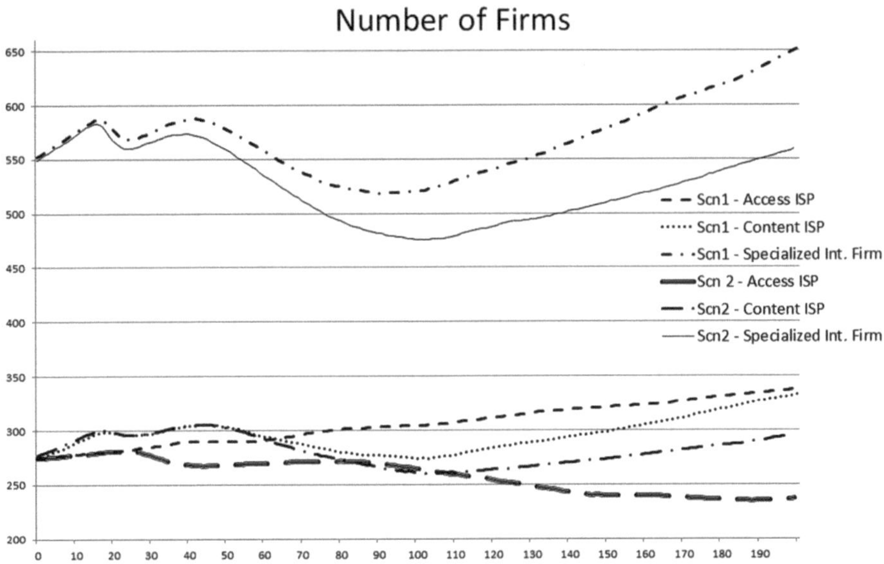

Fig. 2.3 A simulation of the Nordic ISP industry. The timeline represents outcome variables in scenario 1 and 2. Reported values are an average of 50 simulations per scenario

value for the content ISPs). To increase the success threshold value means to force firms to do more R&D in order to reach the increased target value of profit. Otherwise all the parameters remain the same.

2.7 Empirical Results from Simulation of Two Scenarios

In our simulation we study the development in our scenarios resembling the Nordic ISP industry. As we pointed out above scenarios are stylized and based on ideal-typical firms, but Scenario 1 is designed to mimic certain empirical facts of the industry. Scenario 2 is counterfactual simulation of what could have happened if the large incumbent access ISPs were to change their innovation and partnering strategy in a more outward-oriented exploratory direction. The simulation was run 50 times for each of the scenarios. The graph statistics and Fig. 2.3 are an average of those simulations. Figures 2.4 and 2.5 are not an average visualization of the 50 simulations. However, Figs. 2.4 and 2.5 are a visualization of the simulation with values closest to the average graph statistics for the 50 simulations for each of the scenarios. In this way the two specific simulations shown are considered a representative visualization. These figures are an aggregate of the 200 time periods and are used to display differences between the scenarios.

Our first research question asked how the firm-level adoption of specific innovation strategies affects the structural evolution of innovation networks within the

Fig. 2.4 Industrial knowledge-base dynamics—scenario 1

context of a single knowledge intensive industry. In the following we discuss and respond to this question.

2.7.1 The Evolution of Innovation Networks

When looking at the evolution of innovation networks, we were particularly interested in how a change in the propensity of large access IPS to form network ties with external firms affects the structure of ties among actors in the whole network.

The cluster coefficient is 29% lower in scenario 1 than in scenario 2. This means the ties among firms in the market are more dispersed in scenario 1 compared to scenario 2. The interpretation is that scenario 1 has a more heterogeneous inter-firm collaboration pattern. In other words, there are more and smaller local clusters

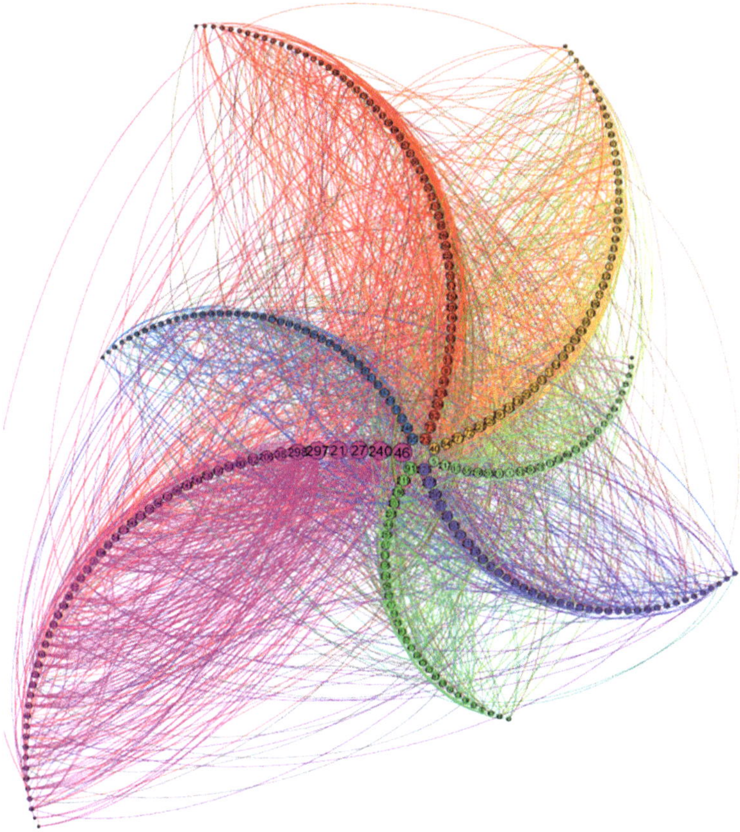

Fig. 2.5 Industrial knowledge base dynamics—scenario 2

(communities) of firms with interlinking ties in scenario 1. Scenario 2 has fewer local clusters with fewer ties between the clusters.

This also implies a more homogeneous market in scenario 2, in terms of inter-firm collaboration patterns in the market. To back up this assertion based on the cluster coefficient, the number of modular communities is 32% higher and triangles is 5% higher, which also confirms the more heterogeneous market in scenario 1. The number of modular communities and triangles means that there are more local clusters in scenario 1 than scenario 2. It is a market environment where many firms operate in smaller local clusters. Reading the reported values in Table 2.3, scenario 2 represents a market with fewer local clusters of inter-firm collaboration that are less dispersed.

The degree and the weighted degree tell us an interesting story of connectedness combined with strong and weak ties in the market. The 3.5% increase in degree from scenario 1 to scenario 2 means the firms are more connected via ties. However, the 7% decrease in weighted degree means that each of the partnerships are not on

Table 2.3 A comparison of social network analysis estimates for scenario 1 and 2. All reported values represent an average of 50 simulations

Scenario firm statistics	Degree	Weighted degree	Cluster coefficient	Modular coefficient (Q)	Modular communities	Triangles
Scenario 1	5.5915	175.46898	0.0409	0.56	300	2,901.6
Scenario 2	5.78756	162.7286	0.05786	0.52	227.48	2,756.66

average re-established as often in scenario 2. These two differences in percentage are not so big in themselves; however, they are large enough to constitute a change in the way the agents are connected. Thus we have a situation where there are more partners, but weaker ties with each partner in scenario 2. In the model successful firms do not need to initiate partnerships. A likely conclusion is that the firms who survive are more successful in scenario 2. Thus a shift from inward-oriented and exploitative to outward-oriented and explorative business strategies induces a more homogeneous market in terms of inter-firm collaboration, the ties become weaker and more frequent. In other words, scenario 2 has more ties within each community, but they are weaker. Instead of dispersed local clusters (the heterogeneity in scenario 1), the firms are more clustered in scenario 2, giving a more uniform inter-firm collaboration pattern.

In this section we have mostly dealt with what happens to the firms and the market structure by analyzing aggregated industry-level data and performing a social network analysis. In the next part we shall however study consequences for firms' economic performance.

2.7.2 *Firms Economic Performance*

How does the evolution of different kinds of network structures affect the economic performance of individual firms populating the industry?

Scenario 1 is calibrated to mimic certain features of the evolution in the Nordic ISP industry the last 15 years over 200 time periods. As such we interpret Fig. 2.3 as a representation of the development over a period of time roughly equal to 15 years, which makes one time period roughly the equivalent of 1 month. The lines representing scenario 1 reflect that the Internet service industry in general is growing. All of the firms, disregarding differences in firm type, are in competition with each other, since they all have the capacity to independently trade Internet access capacity and associated services. This, however, does not mean that all firms are competing over the same customers at all times. The SIS firms share their capabilities employed in the kenes forming their services with the access and content ISP firms. As such they are overly exposed to competition from access and content ISP firms. As the simulation settles after the initial 25–30 time periods, the SIS firms loose in competition and many firms die. Half way through they manage to adapt to the competition and more SIS firms get established and survive.

Some of the decline in SIS firms after 20–40 periods is due to initial capital running out. The same is true for content ISP firms, but together with access ISP firms they manage to survive the competition as they have more capabilities to choose from. Their services are more diversified. The access ISP firms have more initial capital and do not show the same mortality rate as the content ISP firms. The presence of a few large access ISP firms seems to give them an advantage over the content ISP firms.

Scenario 2 is identical to scenario 1, save a change in the parameter for cooperation and internal R&D in access ISP firms. Instead of being conservative in their choice of partner, they now extend their search for possible partners to firms with a more different kene configuration and they increase their own R&D effort. This amounts to what we referred to above as a more outward-oriented and explorative strategy and increases the sharing of knowledge in the market. The immediate result for access ISPs is that they start to compete more amongst themselves and there is a strong decline in the number of access ISPs. With more knowledge shared, the services become increasingly similar, the market more homogeneous, the competition more fierce and fewer firms survive. As a consequence the market conditions for the content ISPs and SIS firms are impacted in a negative manner as well. The overall increase in competition allows fewer content ISPs and SIS firms to survive. However, the change of business strategy in scenario 2 allows the content ISP firms, which are smaller and more innovative, to outperform the access ISPs. There was a trend towards the content ISPs catching up with access ISPs in scenario 1, but with a stronger sharing of knowledge in scenario 2 the content ISP firms catch up much faster.

2.7.3 Distribution of Competencies and Novelty-Creation Capacity

How does the evolution of different kinds of network structures affect the distribution of competencies and the novelty-creation capacity of the industry as a whole?

Our third question addresses how micro-level changes in firms' innovation collaboration strategies affect the structural evolution of knowledge bases within single industries. We are no longer looking into the composition of collaborative ties in the industry, but the composition of knowledge in the industry. In relation to this, we were particularly interested in how a change in the propensity of large access ISPs to form network ties with external firms affects the heterogeneity and interlinking of different forms of knowledge. Our treatment of this question constitutes the main contribution of the paper to existing literature as few contributors have so far analyzed (from an ABM perspective) the dynamics of industrial knowledge bases.

In the last two sections, specifically Table 2.3, we treated firms as nodes and inter-firm collaboration ties as edges. This was done to provide social network statistics describing collaboration structures in the market. The object of analysis has now changed. In this section the term 'node' (round dot in the figures) and 'edge' (linking the nodes) change their meaning. We now refer to a node (the round dots in Figs. 2.4 and 2.5) as a capability, or knowledge area. Two nodes linked by an edge (one of many lines in the figures) means the two capabilities the nodes represent are used in the same service. Figures 2.4 and 2.5 and Table 2.4 can at first be confusing. Figures 2.4 and 2.5 is a visual representation of how particular areas of knowledge is connected and combined in services provided by the firms. We recall a service (or Innovation Hypothesis) is made up from capabilities (particular piece of knowledge). Each node/round dot in Figs. 2.4 and 2.5 represent one of the 300 different capabilities employed in the model and so it is 300 nodes displayed in the figures. A sub-network of capabilities constitutes a service. In other words, a triangle of three round dots in the two figures that are interlinked would be a service provided by a firm at the end of a simulation run. In Figs. 2.4 and 2.5 clusters of nodes are represented as a string of nodes, constituting an "arm". Nodes grouped together in a string (cluster) means that these capabilities have many edges between them and are often used in the same service. As such, we can say that Figs. 2.4 and 2.5 display the distribution of knowledge areas at the end of the simulation for the two representative scenarios. In this way we are looking at a map of the composition of knowledge in the model.

These figures present an overview of how the 300 capabilities are employed at the end of the representative simulations as mentioned above. This particular layout of the figures is chosen to best visualize the number of clusters in the knowledge base. A service built from three kenes has three capabilities displayed as nodes in Figs. 2.4 and 2.5. These three nodes would be connected together via edges and most likely they would belong to the same cluster. However, there are also some edges which bridge the knowledge clusters. The more popular a capability is (higher degree) in conjunction with other capabilities, the closer it is to the center and the larger the size of the node.

The table gives us quantitative information about the two figures. The degree in Figs. 2.4 and 2.5 are roughly the same. This means that the number of capabilities that constitutes a service is about the same, and in turn that the way services are made by firms is not influenced by inter-firm collaboration or innovation strategies. The number of capabilities in a service is then not particularly affected by the change of strategy in scenario 2.

When access ISPs change their strategy in scenario 2 and begin to learn from other firms with knowledge areas more unknown to themselves, we observe a higher degree of clustering of capabilities. More capabilities are linked to each other and used to form services in Fig. 2.5. We see this through the decreased number of clusters (communities, or arms in the figures, made up by knowledge areas) in scenario 2 and the cluster coefficient. This is also the most striking visual difference between Figs. 2.4 and 2.5. The number of knowledge clusters has decreased from 9 in scenario 1 to 7 in scenario 2. With a finite number of knowledge

Table 2.4 A comparison of outcome variables for the knowledge dynamics in scenario 1 (simulation 1 of 50) and scenario 2 (simulation 10 of 50)

Scenario statistics	Degree	Cluster coefficient	Modularity coefficient	Modularity communities	Jaccard index
Scenario 1 (Fig. 2.4)	14.95	0.056	0.247	9	1,016.63
Scenario 2 (Fig. 2.5)	14.45	0.06	0.267	7	1,028.07

areas and a decreased number of clusters, we observe directly in Fig. 2.5 (compared to Fig. 2.4) that the numbers of capabilities present in one cluster increase. Since there are fewer local pockets of knowledge areas used in conjunction with each other, and the distribution of capabilities is more concentrated, the services become more equal each other. With more knowledge areas (equals capabilities or nodes) operating in the same cluster, there is less diversity in the services available in the market. In this way firms are providing more similar services in scenario 2.

This is also consistent with the increased Jaccard index in scenario 2. The Jaccard similarity coefficient is here used to quantify the similarity of one service compared to all the other services provided. The similarity is measured by comparing the difference of capabilities in the services. The Jaccard index is the aggregation of the all the similarity coefficients (300×300 matrix). Alone, the aggregate Jaccard index is only slightly larger in scenario 2 than in scenario 1, but it substantiates our claim that services become more similar, and thus the market more homogeneous in terms of services available in the market, with increased competition in scenario 2. In Sect. 2.6.1 we argued that for scenario 2 the market becomes more homogenous in terms of the distribution pattern of the collaboration ties. Here we argue that scenario 2 is also more homogenous than scenario 1 in its knowledge base.

The model does not answer directly whether more similar services in scenario 2 compared to scenario 1 have a negative impact on the potential for innovation or not. But, drawing on existing research literature about the relationship between innovation network structure, knowledge base heterogeneity and innovation capacity, it is possible to infer some likely implications for innovation. What we have observed is that the change in strategy have led to a stronger competition amongst firms, poorer economic performance, a denser inter-firm collaboration pattern with weaker ties and fewer local collaboration clusters. In addition the knowledge dynamics in the industry has changed with fewer clusters of knowledge areas and more similar services. Drawing on the literature chapter, March (1991) has pointed out the consequence for innovation in a situation like the one present in scenario 2. In his article he points out that repeated interactions between a limited number of firms can lead to homogeneity of knowledge in the network (as is the case in scenario 2), and this in turn generates fewer novel solutions. Thus the change in strategy from scenario 1 to scenario 2 has a negative impact on the potential for innovation.

2.8 Concluding Remarks

It is now time to return to the main question we started out with. How can strategic decisions about collaboration made by a large number of heterogeneous firms affect the structural development of an inter-firm network and an associated network knowledge-base?

To answer this question, we first outlined two-dimensional typology with four distinct innovation collaboration strategies; (1) Exploitative and inward oriented, (2) Explorative and inward oriented, (3) exploitative and outward oriented and (4) explorative and outward oriented. To test how the firm-level adoption of these strategies can affect the development of innovation networks, we developed an ABM that mimics these strategies and some stylized contextual facts of the Nordic ISP industry. In contrast to other modeling exercises (Lave and March 1993; Gilbert 2008), our objective was not to make an exact replication of previous innovation networking activities in this industry. Rather, we built our model on broad stylized facts about the industry environment and innovation strategies that firms have adopted in the past. In the model, we kept macro-level stylized facts of the industry constant while introducing two different innovation collaboration strategies; (a) - inward-oriented and exploitative and (b) outward-oriented and explorative. The purpose was not to explore all possible innovation collaboration strategies and their implications. Rather, we sought to observe what happens in the network and industry knowledge base when we make slight alterations to two specific strategies which have been adopted by a group of firms.

In the model, a change in the collaboration strategies of a sub group of firms had a substantial impact on the evolution of innovation networks and associated knowledge bases. As we pointed out earlier, firms adopting an inward-oriented and exploitative strategy typically carry out their development activities in-house and when they do collaborate, they choose partners with knowledge that is similar to their own. When adjusting the ABM in such a way that many agents follow this strategy, the effects on the macro-level network structure was a relatively low density of linkages between firms, and low density of linkages between knowledge and capabilities. Said differently, the adoption by many firms of inward oriented and exploitative innovation strategies facilitates the emergence of sparsely connected networks with isolated pockets of specialized knowledge. This latter outcome appears to have more beneficial competitive implications for smaller SIS firms that can compete on the basis of their specialized competencies.

By contrast, firms adopting an outward-oriented and explorative startegy seek to develop new knowledge through extensive collaboration and tend to choose partners with knowledge that is different from what they already possess. When adjusting the ABM in such a way that this strategy became more prevalent, the result was the evolution of a relatively tightly interconnected network in which the participant firms build complex but increasingly similar knowledge bases. This seemed to have positive implications for the economic performance of large and wealthy firms that can afford to experiment, but potentially negative effects for smaller specialized firms that are more vulnerable to experimentation. On a macro-level, the interconnection of knowledge resources and the resultant eradication of isolated pockets of knowledge might reduce the network's potential for generating novelty.

On a more general theoretical level, the model results are in line with past ABM analyses on the relationship between firm level strategies and macro-level networks (Gilbert et al. 2001; Ozman 2007, Antonelli and Ferraris 2011), but complements these in one important way. Whereas past studies have typically emphasized the evolving structure of ties between firms populating specific industries, we have added to this research by visualizing also the evolving structure of ties between different strands of knowledge possessed by the firms. Using ABMs to visualizing and analyzing the structure and dynamics of industrial knowledge bases is useful since it allows us to discuss how firm-level decisions about collaboration can affect the novelty-creation capacity of an entire regional industry. One key contribution of this chapter has been to visualize by way of ABM and social network software how firm-level decisions about collaboration affects the evolution of industrial knowledge bases. A possible avenue of future research can be to analyze how variations in the content of industrial knowledge bases can affect the innovative capacity of individual firms.

References

Ahrweiler P, Gilbert N, Pyka A (2011) Agency and structure. A social simulation of knowledge-intensive industries. Comput Math Organ Theor 17:59–76

Ahuja G (2000) Collaboration networks, structural holes, and innovation: a longitudinal study. Adm Sci Q 45(3):425–455

Antonelli C (ed) (2011) Handbook on the economic complexity of technological change. Edward Elgar, Cheltenham, UK

Antonelli C, Ferraris G (2011) Innovation as an emerging system property: an agent based simulation model. J Artif Soc Soc Simulat 14:1–63

Axelrod R (1997) The complexity of cooperation: agent-based models of competition and collaboration. Princeton University Press, Princeton, NJ

Blondel VD, Guillaume JL, Lambiotte R, Lefebvre E (2008) Fast unfolding of communities in large networks. J Stat Mech Theor Exp 2008, P10008

Breschi S, Malerba F, Orsenigo L (2000) Technological regimes and Schumpeterian patterns of innovation. Econ J 110(463):388–410

Burt RS (2004) Structural holes and good ideas. Am J Sociol 110(2):349–399

Chesbrough HW (2003) Open innovation: the new imperative for creating and profiting from technology. Harvard Business School Press, Boston, MA

Chesbrough H, Vanhaverbeke W, West J (2006) Open innovation – researching a new paradigm. Oxford University Press, New York

Dahlander L, Gann DM (2010) How open is innovation? Res Pol 39:699–709

Edquist C (ed) (2003) The Internet and mobile telecommunications system of innovation. Elgar, Cheltenham, UK

Edquist C (2007) The internet and mobile telecommunications system of innovation. Edward Elgar Publishing, Cheltenham, UK

European Network and Information Security Agency (ENISA) (2011) InterX: resilience of the internet interconnection ecosystem. Report April 2011. Downloaded from http://www.enisa.europa.eu/activities/Resilience-and-CIIP/networks-and-services-resilience/inter-x/interx/inter-x

Fjeldstad Ø, Becerra M, Naravan S (2004) Strategic action in network industries: an empirical analysis of the European mobile phone industry. Scand J Manag 20(March–June):173–196

Fortunato S, Barthelemy M (2007) Resolution limit in community detection. Proc Natl Acad Sci U S A 104(1):36

Fraas M, Hauknes J, Smith K, Wibe M, Orstavik F (2001) Corporate governance and innovation in Nordic telecommunications. Report 9 – 2001. STEP Group, Oslo

Gilbert N (2008) Agent-based model. Quantitative applications in the social sciences, vol 153. Sage, Thousand Oaks, CA

Gilbert N, Pyka A, Ahrweiler P (2001) Innovation networks – a simulation approach. J Artif Soc Soc Simulat 4(3):8, downloaded from http://jasss.soc.surrey.ac.uk/4/3/8.html

Gilbert N, Ahrweiler P, Pyka A (2007) Learning in innovation networks: some simulation experiments. Phys A 378:100–109

Gilbert N, Ahrweiler P, Pyka A (2010) The SKIN (simulating knowledge dynamics in innovation networks) model. University of Surrey, University College Dublin and University of Hohenheim

Grandori A, Soda G (1995) Inter-firm networks: antecedents, mechanisms and forms. Organ Stud 16(2):183–214

Hagedoorn J, Schakenraad J (1994) The effect of strategic technology alliances on company performance. Strat Manag J 15:291–309

Hansen MT (1999) The Search-Transfer problem: the role of weak ties in sharing knowledge across organization subunits. Adm Sci Q 44:82–111

Hauknes J, Smith K (2002) Corporate governance and innovation in mobile telecommunications: how did the Nordic area become a world leader? Report R-12 – 2002. STEP Group, Oslo

Jansson J (2011) Emerging (Internet) industry and agglomeration: internet entrepreneurs coping with uncertainty. Entrepren Reg Dev 23(7–8):499–521

Lave J, March JG (1993) An introduction to models in the social sciences. University Press of America, San Francisco, CA

Lechler T, Taylor J, Klingenberg B (2007) The telecommunications carriers' dilemma: innovation vs. network operation. In: Kocaoglu DF, Anderson T, Daim TU (eds) Conference proceedings PICMET 07: management of converging technologies, Portland, OR, vol 6, pp 2940–2947

Lozano S, Arenas A (2007) A model to test how diversity affects resilience in regional innovation networks. J Artif Soc Soc Simulat 10(4):8

Malerba F, Orsenigo L (1996) Schumpeterian patterns of innovation are technology-specific. Res Pol 25:451–478

March JG (1991) Exploration and exploitation in organizational learning. Organ Sci 2(1):71–87

McEvily B, Zaheer A (1999) Bridging ties: a source of firm heterogeneity in competitive capabilities. Strat Manag J 20(12):1133–1156

McGregor T, Alcock S, Karrenberg D (2010) The RIPE NCC Internet data measurement repository. Lect Notes Comput Sci 6032:111–120

Noteboom B (1999) Inter-firm alliances: analysis and design. Routledge, New York

Noteboom B (2002) Trust. forms, foundations, functions, failures and figures. Elgar, Cheltenham, UK

Ozman M (2007) Network formation and strategic firm behaviour to explore and exploit. J Artif Soc Soc Simulat 11(1):7, downloaded from http://jasss.soc.surrey.ac.uk/11/1/7.html

Powell WW, Koput KW, Smith-Doerr L (1996) Interorganizational collaboration and the locus of innovation: networks of learning in biotechnology. Adm Sci Q 41:116–145

Powell WW, White DG, Koput KW, Smith JO (2005) Network dynamics and field evolution: the growth of interorganizational collaboration in the life sciences. Am J Sociol 110(4):1132–1205

Pyka A, Gilbert N, Ahrweiler P (2007) Simulating knowledge-generation and distribution processes in innovation collaborations and networks. Cybern Syst 38(7):667–693

Rodan S, Galunic C (2004) More than network structure: how knowledge heterogeneity influences managerial performance and innovativeness. Strat Manag J 25(6):541–562

Schilling C, Phelps M (2007) Interfirm collaboration networks: the impact of large-scale network structure on firm innovation. Manag Sci 53(7):1113–1126

Sjölander S, Magnusson M, Andrén L (2001) Evolution, adaptation and entrepreneurial learning in the emerging Nordic wireless internet industry. Working paper 2001:5. Department of Innovation Engineering and Management, Chalmers University, Stockholm, Sweden

Taber CS, Timpone RJ (1996) Computational modeling. Quantitative applications in the social sciences, vol 113. Sage University Press, Thousand Oaks, CA

Teece DJ (1986) Profiting from technological innovation: implications for integration, collaboration, licensing and public policy. Res Pol 15(6):285–305

Teece D (1992) Competition, cooperation, and innovation: organizational arrangements for regimes of rapid technological progress. J Econ Behav Organ 18(1):1–25

Uzzi B (1997) Social structure and competition in interfirm networks: the paradox of embeddedness. Adm Sci Q 42:35–67

Zirulia L (2009) The dynamics of networks and the evolution of industries: a survey of the empirical literature, Ch. 3. In: Malerba F, Vonortas NS (eds) Innovation networks in industries. Edward Elgar, Cheltenham, UK, pp 45–77

Chapter 3
The Evaluation of Value Chain Marketing Strategies: An Agent-Based Approach

Stephanie Hintze and Christian Lüthje

Abstract Value chain marketing (VCM) has shown to be a promising strategy to overcome immediate customers' innovation resistance and thus increase the success of supplier innovations. By pursuing VCM, suppliers of entering goods such as raw materials, parts, or components enlarge their target group beyond their immediate customers (manufacturers) and target their downstream customers (applicators) as well. The nature of the VCM process depends on the timing of integrating the immediate customer. Based on the results of multiple case studies focusing on supplier innovations in the field of coatings and sealants, we confirmed that the effectiveness of VCM strategies depends on the newness of supplier innovation and the overlap between the knowledge bases of the involved actors. The analysis further suggested that the newness of innovation and the knowledge overlap are somehow interrelated. In order to assess the overall performance of VCM strategies, we propose an agent-based model and consider a multiplicity of VCM scenarios. In this paper, we provide a theoretical and conceptual foundation for the VCM model to simulate different settings in which supplier innovations are either implemented successfully or fail. We focus on the development of the model and not on the communication of the results of this model.

3.1 Introduction

Firms supplying products to other firms innovate in order to stay competitive in the global market and enhance their sales (see Cowan and Jonard 2009; Kashani 2006). However, they are often faced with big difficulties in marketing, i.e. a high

S. Hintze (✉) • C. Lüthje
Institut für Innovationsmarketing (W-3), Technische Universität Hamburg-Harburg, Schwarzenbergstraße 95, 21073 Hamburg, Germany
e-mail: toth@tuhh.de; c.luethje@tuhh.de

N. Gilbert et al. (eds.), *Simulating Knowledge Dynamics in Innovation Networks*, Understanding Complex Systems, DOI 10.1007/978-3-662-43508-3_3,

resistance to their innovations (Jeannet 2006; Ram and Sheth 1989). This situation is typical for business-to-business (B2B) markets. In these markets, suppliers are confronted with the following problem: Demand for their industrial goods is ultimately derived from demand for the customers' products (Hillebrand and Biemans 2005, 2011; Kleinaltenkamp et al. 2007; Fern and Brown 1984). Due to suppliers' dependence on derived demand, their innovations have to be canalized through many stages in the value chain and need to be accepted and forwarded by many firms in the downstream direction.

To date, suppliers usually approach the value chain by dealing with their immediate customers and essentially pushing their innovations into the value chain (Kleinaltenkamp et al. 2007). But immediate customers often have low incentives to adopt supplier innovations. The value associated with a supplier innovation is not always evident for suppliers' immediate customers. It becomes more obvious when entering goods get closer to their final application. Consequently, immediate customers often prefer to wait until they receive strong signals from their customers indicating the need for an innovation.

To break through immediate customers' innovation resistance, a small number of suppliers try to pursue VCM. They understand that "this narrow view, focusing on the nearest set of buyers, is insufficient for sustained competitive advantage" (Jeannet 2006, p. 14). Therefore, suppliers enlarge their target group beyond their immediate customers and address their downstream customers as well. This wider 'customer horizon' (Hillebrand and Biemans 2011, p. 73) is based on the assumption that downstream customers greatly impact the supplier's marketing success since they have both the potential to benefit from supplier innovations and the market power to pull these innovations through the value chain.

Until now, the VCM literature only focused on the relevance of the topic but did not consider the VCM process in detail. *Pilot interviews* with suppliers in the coating industry helped understand the general VCM process. The nature of this process depends on the timing of integrating the manufacturer. Following the results of the pilot study, we characterize *non-cooperative VCM* as a late integration of the manufacturer. It implies that the manufacturer only takes a passive part in the process. The supplier directly addresses the applicator and the applicator convinces the manufacturer of adopting the innovation without previously consulting him. In contrast, *cooperative VCM* is characterized by an early integration of the manufacturer and necessitates his active participation. The supplier and the manufacturer jointly target the applicator. Pilot interviews further suggested that there is a preference for non-cooperative VCM. This strategy allows suppliers to leapfrog manufacturers and thus reduce their dependence on derived demand.

Based on the pilot study results, *multiple case studies* were conducted with the purpose to deeply analyze the different VCM strategies and explore the variables that seem to affect the supplier's marketing success. The unit of analysis was the supplier's marketing attempt to implement an innovation via VCM. In each respondent firm, a recently finalized marketing project is selected and respondents are asked to discuss the VCM steps and the critical factors for the marketing success. The results showed that the effectiveness of cooperative and non-cooperative VCM depends on two underlying factors: (1) the newness of supplier innovation and (2) the overlap between the knowledge bases of the involved value-chain actors. The analysis of the five cases further suggested that the newness of innovation and the knowledge overlap are somehow interrelated.

With the help of an *agent-based simulation study*, the qualitative results of the case study should be supported. But the central aim of the simulation is to assist in identifying causal relationships that have previously gone unexplained, i.e. the interaction effects between the newness of innovation and the VCM strategy as well as the knowledge overlap and the VCM strategy. In this paper, we provide a theoretical and conceptual foundation for the model of VCM to simulate different settings in which supplier innovations are either implemented successfully or fail.

We structure the paper as follows. First, a theoretical background on VCM is given. We explain cooperative and non-cooperative VCM and the effectiveness of these approaches. Second, we concentrate on the description of the method which includes details on agent-based modeling as an approach in marketing. The review of the literature and the description of the method lead to the conceptual model of VCM. Finally, the findings are summarized and an outlook is provided.

3.2 Value Chain Marketing

Due to suppliers' dependence on derived demand, their innovations have to be canalized through many stages in the value chain and need to be accepted and forwarded by many firms in the downstream direction. Therefore, a perfect understanding of the value chain is one of the challenges suppliers have to meet. Reinforcing this point, Hillebrand and Biemans (2005, p. 5) note that "truly market-oriented firms have a thorough knowledge of the entire value chain." Our understanding of a value chain follows the explanation of Jeannet. He describes the value chain as a system which "includes all industry participants, connected in a successive chain of added value, from raw material production to OEM customers, wholesalers, retail customers and in some cases recycling" (Jeannet 2006, p. 23).

In defining the relevant stages of a value chain, a fixed point of origin and an endpoint are selected (Grunert et al. 2005). The point of departure is represented by a supplier who develops innovative materials. Next, the intermediate stage, i.e. the manufacturer, transforms these materials into a product and sells it to the endpoint, i.e. the stage of the applicator. At this stage, the value of the innovation becomes obvious. The stage of the manufacturer represents the supplier's immediate customer and that of the applicator the supplier's downstream customer.

Jeannet (2006) also explains that VCM goes beyond traditional marketing, missionary selling, and primary demand stimulation. It represents a holistic marketing concept which covers the whole marketing mix (product, price, place, and promotion) and encounters the complexities of the value chain in which a firm operates. Jeannet (2006, p. 22) uses the term VCM "to refer to the practice of influencing an entire industry value chain for the benefit of the marketing function".

The target is to develop comprehensive market intelligence and to promote innovations across all levels of the chain. To stay competitive in the market, industrial marketers have to cover a broader framework to map the value chain. They "must understand not only the cost and revenue dynamics of its intermediate target buyer firms, but also the cost and revenue dynamics facing the buyers' buyers, from whose demand the demand of the immediate market is derived" (Narver and Slater 1990, p. 21).

3.2.1 VCM Strategies

The cross-case analysis indicated that suppliers have the choice to pursue cooperative or non-cooperative VCM to implement their innovations. As already stated in the pilot interviews, there is a preference for *non-cooperative VCM*. This strategy allows suppliers to leapfrog manufacturers and reduce their dependence on derived demand. If suppliers already have direct access to applicators, they prefer non-cooperative VCM. This is based on the fact that applicators are open to supplier innovations and are prepared to take risks. They are continually searching for opportunities to maintain sustainable competitive advantage and value the supplier as an important source of innovation. In contrast, suppliers decide in favor of *cooperative VCM* if the direct access to applicators is lacking.

The selected VCM strategy decides on the pattern of communication. In case of *cooperative VCM*, the message is sent by the supplier and the manufacturer. Both actors transmit information on the innovation's benefit and its technical feasibility. To gain a common understanding, the manufacturer alters the message into an appropriate message for applicators (Duncan and Moriarty 1998). In case of *non-cooperative VCM*, only the supplier sends the message to the applicator. The supplier tries to transmit an appropriate message to overcome the distance to the applicator. Indeed, communication is more effective between transceivers who share common meanings (see Rogers and Kincaid 1981; Rogers and Bhowmik 1971). The effectiveness of communication, in turn, influences the effectiveness of VCM.

3.2.2 Newness of Innovation

Based on the results of multiple case studies, the effectiveness of VCM strategies depends on two variables: (1) the newness of supplier innovation and (2) the overlap between the knowledge bases of the involved value-chain actors. Previous innovation studies differentiate among innovations that differ in their degree of newness based on: (1) the newness as perceived by the producers and (2) the newness as perceived by the target consumers (Ziamou 1999; Veryzer 1998; Kleinschmidt and Cooper 1991; Gobeli and Brown 1987). For the purpose of our research, the classification of Ziamou (1999) captures the essence of newness best. She distills the most relevant dimensions of newness by focusing on the product functionality and the efforts made to provide this functionality. Transferred to the VCM phenomenon, the first dimension refers to the functionality of a final product and the second dimension describes what a manufacturer has to do to provide the functionality expected by an applicator.

Based on this consideration, the first dimension of newness describes the newness as perceived by the applicator and reflects the product dimension of newness. It incorporates changes in the product functionality. The second dimension refers to the newness as perceived by the manufacturer and reflects the technical dimension of newness. The combination of the product and technical dimension of newness determines how novel the supplier innovation is.

The first type of supplier innovation, labeled *incremental innovation*, refers to minor changes to present products. It offers the same benefits to the applicator and requires no changes in the manufacturing process. The second type of supplier innovation, labeled *functionality-driven innovation*, involves the implementation or trading of a new product with enhanced functionality but does not evoke changes in the manufacturing process. Products are adapted from other fields of application and involve the creation of a new field. The third type of supplier innovation, labeled *process-driven innovation*, comprises the adaptation of the manufacturing process of suppliers' immediate customers. While this type of innovation provides the same product functionality, it can offer other benefits to applicators, e.g. lower production costs, enhanced productivity and product quality, as well as resource efficiency. The fourth type of supplier innovation, labeled *really new product*, offers a new functionality to the relevant applicator but also necessitates changes in the manufacturing process.

3.2.3 Knowledge Overlap

Typically, different actors possess different knowledge bases due to their position in the value chain. In the context of VCM, a distinction between three types of knowledge bases has to be made: science-driven knowledge base, engineering-driven knowledge base, and application-driven knowledge base (Asheim 2007).

The *science-driven knowledge base* is found in value chains where innovations are developed. Important activities in this context are basic and applied research as well as systematic development of products and processes (Asheim 2007). Suppliers operate in a science-based area. They have special know-how in the field of chemical composition, structure, and properties of substances and ingredients. To create innovative materials, they typically have their own R&D department but also cooperate with universities and other research organizations.

The *engineering-driven knowledge base* is found in value chains where innovations are converted. Regarding the VCM phenomenon, manufacturers are characterized by an engineering-driven knowledge base. They are experts in the field of formulation and conversion of materials, and focus on the physical manufacturing process and the testing of final solutions.

The *application-driven knowledge base* is found in value chains where the value of supplier innovations becomes obvious. With respect to VCM, applicators are characterized by an application-driven knowledge base. They are mainly oriented toward a deep understanding of their application field to fulfill end user needs and expose market trends. They specialize in the domain of functionality and aesthetics of final products.

To communicate successfully, the actors try to acquire knowledge in other domains and thus induce an overlap between their knowledge bases. If there is a high knowledge overlap in the domains relevant to a message, the communication will be smooth and face less resistance (Fussell and Krauss 1989). The relevance of the actors' knowledge bases and thereby the information embedded in the messages is determined by the newness of the supplier innovation marketed. Functionality-driven innovations call for product-related information. To transmit this kind of information, suppliers should be familiar with the applicator's business. Process-driven innovations, in turn, require technical information. Accordingly, suppliers must understand the manufacturing process. The marketing of really new products entails product-related as well as technical information. As a result, suppliers need to know the manufacturer's and the applicator's business.

3.2.4 Interaction of Newness of Innovation and Knowledge Overlap

To identify different patterns in the use of VCM, each case was examined in relation to the newness of innovation, the knowledge overlap, and the strategy.

The *first pattern* describes a marketing attempt where a supplier tries to implement a functionality-driven innovation via non-cooperative VCM. This type of innovation involves the implementation of a new product with enhanced functionality but does not require changes in the manufacturing process. In two of the five cases, the applicator articulates an urgent need. To correspond to the applicator's need, the supplier reacts by absorbing an existing product from one application field and transmits it to a new field. Due to familiar contact persons, the supplier directly targets the applicator and thus leapfrogs the manufacturer. Despite the distance in the value chain they communicate effectively due to the high knowledge overlap. There are no problems in communication since the supplier and the applicator have high expertise in communicating with each other. Finally, the supplier implements the present innovation successfully. A slight variation of the first pattern can be observed in another case. Here, the final applicator does not formulate an urgent need. Nevertheless, the supplier proactively creates a new product to set new standards in the application field and correspond to future market trends. All other parameters (newness of innovation, knowledge overlap) have the same value. In all three cases, non-cooperative VCM is proven to be the right strategy.

The *second pattern* describes a marketing attempt where one supplier tries to implement a functionality-driven innovation via cooperative VCM. The problem here is that the manufacturer is either unwilling or unable to aggressively promote the supplier innovation that is not directly required by the applicator. He fails to contact the OEM. As a result, the supplier innovation remains with the manufacturer because he prefers to wait until he receives strong signals from OEMs indicating the need for the said innovation.

The *third pattern* characterizes a marketing attempt where a supplier tries to implement a really new product via cooperative VCM. This type of innovation offers a new functionality to the relevant applicator but also requires changes in the manufacturing process. The applicator is proactively searching for an innovative solution to enable the functionality needed. As the supplier has no familiar contact person at the stage of the applicator, he pursues cooperative VCM. The supplier first addresses the manufacturer to gain his support because he has an existing relationship with the applicator. Next, both actors jointly target the applicator to present the innovation. Due to the small knowledge overlap between the supplier and the applicator, the manufacturer acts as a mediator. This implies that the manufacturer reduces the big distance between them and eliminates problems in communication. In this case, cooperative VCM is the right marketing strategy to implement the supplier innovation successfully.

The *fourth pattern* describes a marketing attempt where a supplier tries to implement a process-driven innovation via non-cooperative VCM. This type of innovation requires changes in the manufacturing process. While it provides more or less the same product functionality, it offers other benefits to the applicator such as sustainability. The applicator does not have an urgent need. He just formulates a general goal. Here, the supplier proactively develops a new product. Both the supplier and the applicator communicate effectively because of a high knowledge overlap. Furthermore, they have high expertise in communicating with each other. Finally, the supplier implements the innovation successfully.

3.3 Agent-Based Modeling in Marketing

Until now, the opportunities of using ABM in the field of marketing have been realized by some researchers, but a widespread acceptance and publication of this method cannot be observed. Recently, there has been practitioners interest in using agent-based modeling to explore marketing mix models (Garcia and Jager 2011; Rand and Rust 2011).

One existing ABM in the field is the model of *Garcia and Atkin* (2005). Here, the aim is to gain insights into how coopetition strategies (through a coordinated marketing campaign) can affect the diffusion of resistant innovations. Another existing ABM in this field is a model to study the effectiveness of different promotional strategies (*Delre et al.* 2007). This model in particular focuses on the targeting and timing of the promotions. *North et al.* (2010), in turn, develop a model to study a multi-scale consumer market. This model simulates the interactions between consumers, retailers, and manufacturers of consumer packaged goods.

The benefits of ABM over other simulation methods are numerous. First, the unit of study is the individual or agent. Each agent has unique characteristics and decision rules it follows. It allows for a more realistic representation of the relevant phenomenon (Garcia 2005). With respect to our research, numerous agents with different knowledge bases can be created. Second, the behavior of a single agent (micro-level) represents the basis for the emergence of a collective's behavior as a whole (macro-level). Therefore, agent-based models can be used to illustrate systematically how simple rules of micro-level interaction lead to global patterns or macro-level phenomena (Janssen and Ostrom 2006). This benefit is based on the bottom-up approach by which an ABM is constructed. The programmer of an agent-based model only models the behavior and objectives of an individual because he or she cannot understand the whole phenomenon of interest. These individuals then interact and communicate with each other in a repetitive process. The repeated interactions, in turn, lead to global or macro trends and behaviors (Garcia 2005). In the VCM model, the supplier interacts with applicators and/or manufacturers to implement an innovation. Third, ABM is one among very few methods that enables adaptation or learning process on a microscopic level. Here, suppliers get the ability to adapt their parameters of action over time. Fourth, using ABM does not require an understanding of differential equations, integrals, or even statistics. It is easier to develop than other analytical models (Schramm et al. 2010). This advantage facilitates the development of the VCM model.

In our research, ABM is used in the context of marketing and innovation. The basic concept here is to model the acceptance and implementation of supplier innovations by first describing simple rules of behavior for the types of agents (suppliers, manufacturers, and applicators) and then aggregating these rules. In particular, the case study results are used to set the agents' rules of behavior. The aim is to analyze and assess a multiplicity of VCM settings while looking at the reactions they provoke in a population. It helps to confirm the case study results and assist in identifying causal relationships that have previously gone unexplained (Garcia 2005). Finally, the VCM model is intended as a tool for suppliers to evaluate different VCM strategies in advance.

3.4 The Conceptual Model

Motivated by the SKIN model of Gilbert, Ahrweiler and Pyka, the model presented here includes three types of agents comprising suppliers, manufacturers, and applicators (see Ahrweiler et al. 2004; Gilbert et al. 2001). The agent-based model starts with the individual decision making of the supplier. He tries to promote innovations (functionality driven, process driven, or really new) via cooperative or non-cooperative VCM. Next, the supplier chooses one manufacturer and/or applicator as target of his marketing attempt. The performance of a supplier's attempt is measured in terms of the acceptance and implementation of an innovation as well as the marketing resources used. The basic VCM model is extended with a representation of knowledge dynamics in and between the firms. In particular, the supplier tries to increase the marketing success by improving his knowledge base through adaptation to applicators' needs and learning.

In the next paragraphs, the elements and processes of the model are described in further detail. Existing theories and results of studies are used as the grounds in building the model. Some assumptions are specific to the coatings and sealants industry, following our results of the case study. Still, the model can be extended or adapted to be used in a wider scope. The description of this conceptual model will follow the protocol developed by Grimm et al. (2006, 2010).

3.4.1 *Purpose*

The first goal of this model is to compare the marketing performance of cooperative and non-cooperative VCM. The performance is measured based on the effectiveness and efficiency. Effectiveness represents the rate of successfully implemented supplier innovations, whereas efficiency is related to money and time a supplier spends to perform his marketing attempt (Sheth and Sisodia 2002).

The second objective is to systematically study the effect of the newness of innovation and the knowledge overlap through a relative comparison of the marketing performance of both VCM strategies.

3.4.2 *Entities, State Variables, and Scales*

In the VCM process, there are different actors playing distinct roles. These actors are represented as agents and are conceptualized as heterogeneous agents with respect to their perceptions, actions, and particular attributes. They interact within the value chain. An agent uses its knowledge to interact according to its behavior rules in order to reach its goal. The goal of the supplier is to promote his innovation. Therefore, the supplier gets the ability to adapt his parameters of action over time. The basic variable of each agent is its kene set.

3.4.2.1 Kenes

Each agent is specialized and starts the simulation with specific attribute values and with different knowledge bases. In the VCM model, the representation of the agents' knowledge base draws on the concept of 'kene' developed by Ahrweiler

Fig. 3.1 A firm's kene set

$$\begin{bmatrix} K_0 \\ A_0 \\ E_0 \end{bmatrix} \begin{bmatrix} K_1 \\ A_1 \\ E_1 \end{bmatrix} \begin{bmatrix} K_2 \\ A_2 \\ E_2 \end{bmatrix} \cdots \begin{bmatrix} K_7 \\ A_7 \\ E_7 \end{bmatrix}$$

Table 3.1 Specification of the relevant agent's fields of knowledge

K	Fields of knowledge
1	Chemical properties of substances and ingredients
2	Chemical composition and transformation of materials
3	Formulation and testing of solutions
4	Converting and finishing of solutions
5	Functionality and aesthetics of final products
6	Market trends and regulations

et al. (2004). A kene consists of a vector containing different units of knowledge called triples. Each triple is characterized by three different elements: *K*, *A*, and *E*. The *K* in the VCM model describes the knowledge field. It represents an area in which a firm has specialized knowledge (e.g. chemical composition and transformation of materials). The second element, labeled *A*, refers to the amount of knowledge a firm has achieved in a specific knowledge field. The third element, labeled *E*, represents the expertise which reflects a firm's expertise gathered in communicating its knowledge across the respective value chain. This element is integrated to take the communication aspect of the VCM process into account.

In the VCM model, a firm's kene set consists of eight triples. A real firm might have more units of knowledge than eight triples. But a standard amount is applied to represent the small range of knowledge fields that are appropriate for specific innovations like supplier innovations. Also, it shapes up as an adequate knowledge space size which reports stable results. The focus here is on the composition of a firm's knowledge set, thereby the assumption is made that all actors have the same size of their respective knowledge set (Conti and Hoisl 2012). Figure 3.1 visualizes the kene set of a firm.

The first kene component, the knowledge fields, refers to the breadth of knowledge a firm has (see Prabhu et al. 2005). Dosi et al. (2003) define organizational knowledge as the ability of an organization to perform its characteristic actions in order to develop products. These fields include the series of knowledge required to convert materials into final products (see Table 3.1). The *Ks* are sorted starting with those most related to suppliers placed in the upstream position of a value chain and ending with those most related to applicators placed in the downstream position of a value chain. An integer from 1 to 6 is assigned to each *K* that will be used in the mapping or calculation throughout the model.

The selection of knowledge fields is based on pilot and case study interviews as well as papers that deal with the coatings and sealants industry. Accordingly, the fields are divided into three different categories: chemical (*K* 1–2), technical (*K* 3–4), and application-related knowledge fields (*K* 5–6).

The second kene element, the amount of within-field knowledge, refers to the depth of understanding a firm has with regard to a specific knowledge field (Prabhu

Table 3.2 Specification of the agent's amounts of within-field knowledge

A	Amounts of within-field knowledge
1	Small amount of within-field knowledge (basic knowledge)
2	Medium amount of within-field knowledge
3	High amount of within-field knowledge (deep knowledge)

Table 3.3 Specification of the agent's expertise levels in communication

E	Expertise levels
1	Low expertise level (lack of experience in communication)
2	Medium expertise level
3	High expertise level (high experience in communication)

et al. 2005). It is described by qualitative degrees. Each level of knowledge amount is represented by an integer from 1 to 3 (see Table 3.2). Higher integer symbolizes a deeper understanding of a firm in a certain field. It represents a firm's core competence that is applied in many cases.

The third kene element is the expertise level, which refers to the skill level that a firm has achieved from past experience in communicating its knowledge along the value chain. As mentioned before, it is integrated to consider the communication aspect of the VCM process. Similar to the amount of within-field knowledge, the past experience is ordered into qualitative degrees (see Table 3.3). High degree of expertise implies that an actor has applied its knowledge in different projects.

3.4.2.2 Marketing Concept

Another variable of the agents is their marketing concept. It corresponds to the idea of the innovation hypothesis used in the *SKIN model* where firms apply their knowledge to create innovations (see e.g. Ahrweiler et al. 2004, 2011).

In the VCM model, the marketing concept represents a firm's strategy or philosophy to satisfy the needs of the target market. Therefore, it describes the source a firm uses for its attempts to make profits on the relevant market. The marketing concept is built by two triples that are selected from a firm's kene set. These triples represent different business units or teams in a firm who are working together to address potential business partners. Each supplier and each applicator has his own marketing concept. Still, the idea of this concept is slightly different for these two types of agents. From the supplier's perspective, the concept (MC^S) describes the source he uses for his attempts to implement his innovation along the value chain. On the applicator's side, the concept (MC^A) is the source he uses for his attempts to signalize his needs to direct and indirect suppliers in the value chain.

3.4.2.3 Newness of Innovation

In the VCM model, the newness or type of innovation is derived using the two *Ks* that are available in the supplier's marketing concept (MC^S). As already mentioned, the main goal of the supplier is to satisfy expressed or future needs of customers, especially downstream customers.

If both *Ks* range from 3 to 4, the supplier focuses on technical fields and offers a *process-driven innovation*. If both *Ks* are within 5–6, the supplier focuses on product-related fields and tries to implement a *functionality-driven innovation*. If one of the *Ks* is within 3–4 and the other *K* is within 5–6, the supplier offers a *really new product*.

3.4.2.4 Knowledge Overlap

Concerning the knowledge overlap, the knowledge similarity of two firms is measured. This value is derived from the *Ks* in a firm's kene set. The knowledge overlap equals the number of common *Ks* that belong to the two firms of interest (Conti and Hoisl 2012). In case of non-cooperative VCM, the knowledge fields of the supplier and that of the applicator are compared. If the supplier decides to integrate the manufacturer, the knowledge fields of the supplier and the manufacturer are summed and then compared to the fields of the applicator. This is based on the fact that the manufacturer supports the supplier's attempt to implement a present innovation. The manufacturer can improve the supplier's kene set. He has experience in interacting with downstream customers as it is his daily business.

3.4.2.5 Supplier Size and Marketing Capital

In the model, a supplier is active if he has an innovation and tries to implement it. In order to perform his marketing attempt, a supplier needs to have capital, which refers to the amount of budget that can be used for marketing. Like in reality, the specific amount of capital depends on the firm's size. In the model presented here, big supplier firms have twice as much capital as "normal" supplier firms. If a supplier firm runs out of capital, it cannot perform anymore attempts.

3.4.3 Initialization

At the beginning of each simulation experiment, the first step is the initialization of the agents' kenes which is based on the pilot and case study results. As mentioned before, a kene represents a firm's knowledge base. In fact, it is possible that identical triples appear in a firm's kene. It shows the possibility of having more

Table 3.4 Initial distribution of agent's kene elements

Agent	Kene triple's element	Distribution
Supplier	K	Three of the kene triples are randomly distributed within the range of 1–2
		Three of the kene triples are randomly distributed within the range of 3–6 – If the first of these three Ks is within the range of 3–4, all of these three triples are within the range of 3–4 – If the first of these three Ks is within the range of 5–6, all of these three triples are within the range of 5–6
		Two of the kene triples are randomly distributed within the range of 4–6
	A	For the triples with a K from 1 to 2, the A of those triples is 3
		For the triples with a K from 3 to 6, the A is randomly distributed within the range of 1–3
Manufacturer	K	Four of the kene triples are randomly distributed within the range of 3–4
		Two of the kene triples are randomly distributed within the range of 1–2
		Two of the kene triples are randomly distributed within the range of 5–6
	A	For the triples with a K from 3 to 4, the A of those triples is 3
		For the triples with a K from 1 to 2 or 5 to 6, the A is randomly distributed within the range of 1–3
Applicator	K	Five of the kene triples are randomly distributed within the range of 5–6
		Three of the kene triples are randomly distributed within the range of 1–4
	A	For the triples with a K from 5 to 6, the A of those triples is 3
		For the triples with a K from 1 to 4, the A is randomly distributed within the range of 1–3

than one group in a firm which focuses on the same field of knowledge and has acquired the same amount of knowledge as well as expertise. The initialization rules are summarized in Table 3.4.

For every type of agent, specific rules are defined and different thresholds regarding the distribution of *Ks* and *As* are developed. This is based on the fact that suppliers, manufacturers, and applicators play different roles due to their respective position in a value chain. Suppliers have special know-how in chemical fields, manufacturers are experts in technical fields, and applicators specialize in product-related fields. In other words, every type of agent is characterized by a specific breadth of knowledge (K) and a specific depth of knowledge (A). They are more familiar with their own knowledge field than with surrounding fields. By contrast, the agent's expertise level (E) does not depend on its value-chain position. This implies that *Es* are distributed randomly without following any rules.

During the initialization, the agents' marketing concepts are built from their kene sets. As mentioned before, a marketing concept describes a firm's strategy to

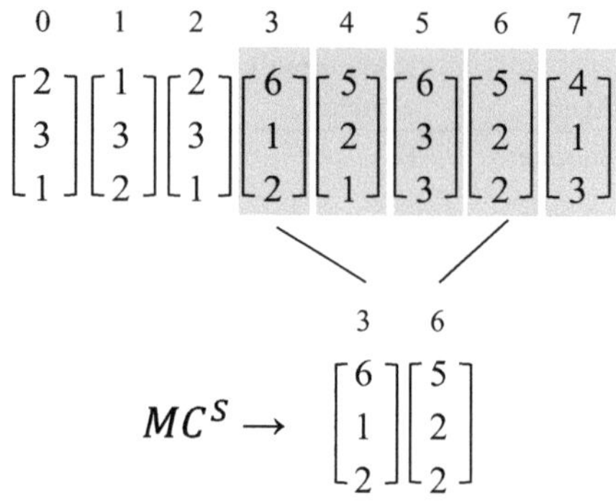

Fig. 3.2 Selection process of a supplier's marketing concept

fulfill customer needs. The goal of the supplier is to satisfy the needs of his immediate and downstream customers. Thus, his marketing concept (MC^S) is composed by two triples that provides technical (K 3 or 4) and/or product-related improvements (K 5 or 6). In other words, K is higher than or equal to 3. In addition, the two triples represent *Ks* which occur most often. This rule is based on the understanding that a firm builds its marketing concept based on the knowledge fields it has focused on. These fields are highly relevant in the value chain in which a firm operates. Figure 3.2 visualizes the formulation of a supplier's marketing concept.

In the presented kene set, it can be seen that triple-3 to -7 (highlighted in gray) are the triples with *Ks* higher than or equal to 3. These are the triples that are qualified to be selected into the supplier's marketing concept because they can provide a new product with enhanced productivity, resource efficiency, and product functionality. Here, K 5 and 6 are the fields which occur more than once. These fields have main importance in the respective value chain. In building a marketing concept, one of the triples with K 5 and one with K 6 are selected randomly. Here, triple-3 and -6 are selected and represent the supplier's marketing concept.

3.4.4 Process Overview and Scheduling

After the initialization, all agents are created and ready to play their roles. The process is started by suppliers. A single process ends if there is an implementation or a rejection of the innovation. If an innovation is rejected, the supplier tries to improve his marketing concept and starts a new attempt. If possible, the supplier searches through the qualified triples ($K \geq 3$) and selects a triple with a higher A.

An example is shown in Fig. 3.3. In the first step, the supplier selects triple-3 and -6 in his marketing concept (highlighted in light gray). Triple-3 is characterized by a low A (1). To be successful in the second step, the supplier searches for a triple with the same K (6) but a higher A (2 or 3). These conditions are fulfilled in triple-5 (highlighted in dark gray). As a result, the new marketing concept of the supplier

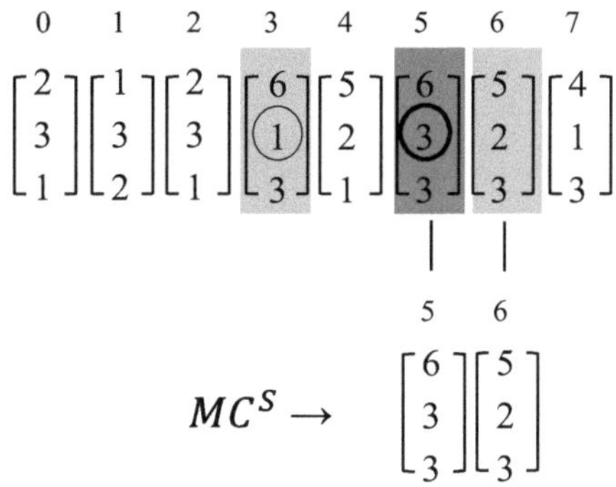

Fig. 3.3 Improvement process of a supplier's marketing concept

consists of triple-5 and -6. Furthermore, *E* of triple-3 and triple-6 has increased by one because these triples have been used in the initial marketing concept and the supplier has obtained learning experience. He gets some feedback from (potential) customers.

The users of the VCM model are given the options to run the process using cooperative and non-cooperative VCM. These two processes are separated to enable a comparison of their effectiveness and efficiency. The process of each strategy is described as follows.

3.4.4.1 Non-cooperative VCM

The process of non-cooperative VCM starts with the supplier selecting one applicator randomly as a target of his marketing attempt. A single non-cooperative process by a single supplier is visualized in a flow chart and shown in Fig. 3.4. A detailed description of each step is given as follows.

Evaluation of the Marketing Objectives

At first, the supplier addresses an applicator to present his innovation and to create awareness for the new idea. The decision of the applicator to invite the supplier to gain further knowledge about the innovation depends on the comparison of the supplier's marketing objectives and his own objectives. As depicted in Eq. (3.1), the applicator compares the knowledge fields (*Ks*) used in the supplier's marketing concept (MC^S) with his own *Ks* used in his marketing concept (MC^A).

$$Invitation: \begin{cases} yes, if\ K_1^S = K_1^A\ or\ K_1^S = K_2^A\ or\ K_2^S = K_1^A\ or\ K_2^S = K_2^A \\ no, if\ K_1^S \neq K_1^A\ or\ K_1^S \neq K_2^A\ or\ K_2^S \neq K_1^A\ or\ K_2^S \neq K_2^A \end{cases} \quad (3.1)$$

As mentioned, the applicator is open to supplier innovations. He does not expect a perfect match as he is interested in new ideas to solve existing and future problems

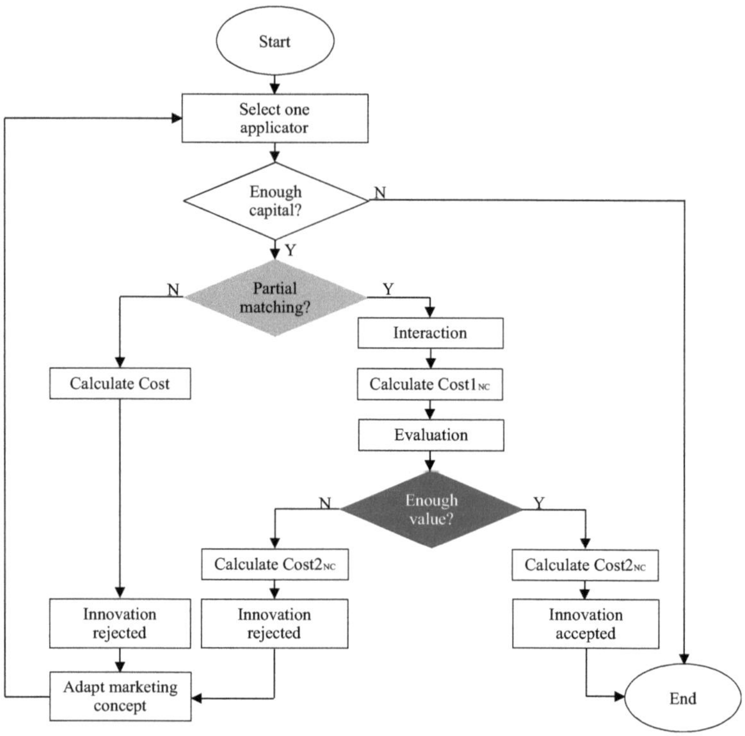

Fig. 3.4 Non-cooperative VCM process

as well as differentiate himself from competitors. Actually, applicators perceive suppliers as an important and attractive source of innovation. This implies that if at least one of the *Ks* is present in both marketing concepts, the applicator develops an interest and invites the supplier. To summarize, the supplier and the applicator must have at least one same first item on the marketing agenda.

If there is no partial match, the supplier information is not relevant to the applicator. Thus, the applicator rejects the innovation. Next, the supplier tries to adapt his concept by replacing one of the selected triples and by drawing another triple with a higher *A* (cf. Fig. 3.3). Furthermore, the *Es* of the used triples increase by one as the supplier learn about the audience to transmit more relevant information.

Interaction Process

After comparing the marketing objectives, the interaction process between a supplier and an applicator starts. Both create a marketing message and transmit it to the other party. To create a message, the agents use their marketing concept. As mentioned before, MC^S describes the source a supplier uses for his attempt to implement his present innovation. Developing this concept into a message (M) is a mapping procedure where the Ks and the As of the marketing concept are used to compute an index number that represents the supplier's message [see Eq. (3.2)].

$$M = \sum_{i \in MC^S} (K_i \times A_i) \tag{3.2}$$

On the other side, the marketing concept of the applicator (MC^A) describes the source he uses to signalize his needs to the supplier. This concept is applied to generate the expected message (EM) of the applicator that is described by

$$EM = \sum_{i \in MC^A} (K_i \times A_i) \tag{3.3}$$

Calculation of the Non-cooperative Marketing Cost 1 ($\mathrm{Cost1}_{\mathrm{NC}}$)

In non-cooperative VCM, the first cost that a supplier spends is the cost to address an applicator in order to transmit his marketing message. If the supplier tries to match M to EM, there is communication going on. Two parties could effectively communicate if they share common knowledge. In this case, the supplier could transmit his message to the applicator easier. If M and EM are totally different, the supplier would need to spend more efforts to interact. Based on this argumentation, the first cost factor in non-cooperative VCM is proportional to the difference of M and EM as depicted in Eq. (3.4). The equation is scaled using a constant ($C1_{NC}$) to enable a comparison with other cost factors. More details on the cost calculation can be found in the Appendix.

$$Cost1_{NC} = (C1_{NC} \times |M - EM|) \tag{3.4}$$

Evaluation Process

After the interaction process, the applicator attempts to judge the value the supplier innovation offers. The value as perceived by the applicator depends on the supplier's marketing performance. The applicator could sense the value of the innovation only if the supplier has a deep understanding regarding certain fields of knowledge and if the supplier is also able to transmit this understanding appropriately to the applicator. Using this argumentation, the customer value as perceived by the applicator is a function of K, A, and E.

The customer value offered by the supplier is derived from his marketing concept (MC^S). It describes the value the supplier innovation offers to the applicator. This value is proportional to his fields of knowledge (K_j) and his expertise (E_j). It can only be created with at least an average amount of within-field knowledge ($A_j \in \{2, 3\}$) [see Eqs. (3.5), (3.6), and (3.7)]. A low amount ($A_j \in \{1\}$) does not contribute to the customer value because it refers to a missing competence of the supplier to speak on a subject.

$$CV_{NC} = \sum_{j \in MC^S} (cv_j) \tag{3.5}$$

where

$$cv_j = K_j \times E_j \quad \text{if} \quad A_j \in \{2, 3\} \quad j \in MC^S \tag{3.6}$$

$$cv_j = 0 \quad \text{if} \quad A_j \in \{1\} \quad j \in MC^S \tag{3.7}$$

Besides the value offered by the supplier, the applicator also has a certain expected or desired value (ECV_{NC}). This expected value is what the applicator looks for in order to achieve his main goal (Flint et al. 1997). In the model, the expected value is derived from the applicator's marketing concept (MC^A). It is proportional to his fields of knowledge (K_j) and his expertise (E_j). The calculation is similar to the one of the offered value (CV_{NC}) [see Eqs. (3.8), (3.9), and (3.10)].

$$ECV_{NC} = \sum_{j \in MC^A} (ecv_j) \tag{3.8}$$

where

$$ecv_j = K_j \times E_j \quad \text{if} \quad A_j \in \{2, 3\} \quad j \in MC^A \tag{3.9}$$

$$ecv_j = 0 \quad \text{if} \quad A_j \in \{1\} \quad j \in MC^A \tag{3.10}$$

Comparison of the Offered and the Expected Customer Value

At this stage, the applicator wants to know if the value offered by the supplier meets his expected or desired value. In the VCM model, the applicator compares CV_{NC} and ECV_{NC} to make his decision to accept or reject the innovation as formulated in Eq. (3.11). This decision is highly critical in non-cooperative VCM. The applicator accepts the innovation if the value offered by the supplier corresponds or exceeds his expected value. In this case, the applicator gains a favorable attitude toward the supplier innovation. Otherwise, the supplier innovation is rejected.

$$Acceptance: \begin{cases} yes, if\, CV_{NC} \geq ECV_{NC} \\ no, if\, CV_{NC} < ECV_{NC} \end{cases} \tag{3.11}$$

Despite any rejection, the supplier has gone through a full VCM process at this stage. He has interacted with an applicator and has gained some learning (learning-by-doing). Actually, the supplier becomes more practiced and more efficient at doing what he is already doing (see Cohen and Levinthal 1989). Therefore, he has improved his knowledge base and this shall be reflected in his kene set. The amount of knowledge and the expertise of the supplier are increased by one for the triples that have been used in the marketing concept. This, however, can only happen if there is still room for improvement, i.e. A and E are lower than 3.

Calculation of the Non-cooperative Marketing Cost 2 ($Cost2_{NC}$)

In non-cooperative VCM, the second cost that a supplier spends is the cost to explain and demonstrate the value of his present innovation. This cost element depends on the supplier's marketing performance. Once again, the applicator could evaluate the innovation attributes only if the supplier has a deep knowledge in the relevant fields and if the supplier is able to transmit this knowledge appropriately to the applicator. In the case of supplier's inability, the value offered by the supplier will diverge widely from the value expected by the applicator. Consequently, the interaction process between the supplier and the applicator is characterized by communication problems and thus high costs.

Based on this argumentation the second cost factor in non-cooperative VCM is proportional to the difference of CV_{NC} and ECV_{NC} as depicted in Eq. (3.12). This implies that the bigger the difference between the offered and expected value, the higher the cost the supplier needs to spend in order to convince the applicator. The equation is scaled using a constant ($C2_{NC}$) to enable a comparison to other cost factors. More details on the cost calculation can be found in the Appendix.

$$Cost2_{NC} = (C2_{NC} \times |CV_{NC} - ECV_{NC}|) \quad (3.12)$$

3.4.4.2 Cooperative VCM

After having an overview of non-cooperative VCM, the different steps of cooperative VCM should be described. The special characteristic of this strategy is that the manufacturer is integrated at the beginning of the VCM process. Therefore, the supplier first selects one manufacturer randomly. The single process of cooperative VCM by a single supplier is shown in Fig. 3.5.

Evaluation of the Marketing Objectives

At first, the supplier approaches a manufacturer to ask for support to implement his innovation. By early integration of the manufacturer, the supplier could take advantage of the manufacturer's knowledge. But the decision of the manufacturer to support the supplier's marketing attempt depends on his comparison of the objectives of the supplier and that of the applicator. Due to his position in the value chain, the manufacturer is well-informed on the present objectives of the supplier and the applicator. He only supports a supplier's attempt if both knowledge fields (*Ks*) used in the supplier's concept (MC^S) also occur in the applicator's concept (MC^A) [see Eq. (3.13)]. Thus, he expects a perfect match.

$$Support: \begin{cases} yes, if\ K_1^S = K_1^A\ and\ K_2^S = K_2^A\ or\ K_1^S = K_2^A\ and\ K_2^S = K_1^A \\ no, if\ K_1^S \neq K_1^A\ and\ K_2^S \neq K_2^A\ or\ K_1^S \neq K_2^A\ and\ K_2^S \neq K_1^A \end{cases} \quad (3.13)$$

The manufacturer is characterized by an antagonistic attitude toward supplier innovations. He expects that the focal supplier innovation fully corresponds to the

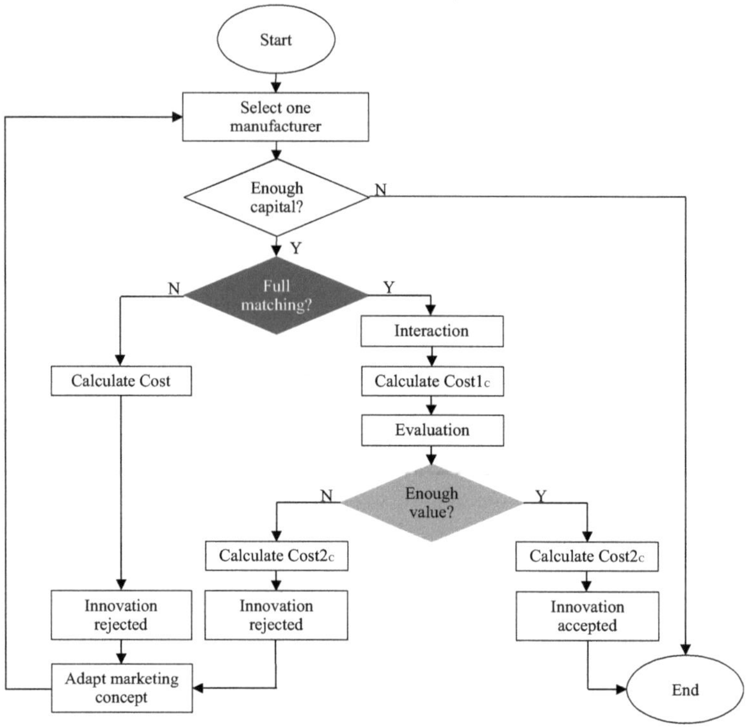

Fig. 3.5 Cooperative VCM process

relevant applicator's needs. This implies that the applicator must require the innovation offered by the supplier. Only in this case, the manufacturer is willing to support the supplier's attempt to not place the business relationship with the applicator at risk. But this threshold is hard to fulfill. If the applicator has one or two other items on his agenda, the manufacturer rejects the innovation. In this case, the manufacturer does not feel impelled to cooperate. In other words, the need uncertainty is too high. Consequently, the supplier has to start a new marketing attempt.

Interaction Process

After comparing the marketing objectives, the interaction process between the supplier, the manufacturer, and the applicator starts. First, the supplier and the manufacturer create a joint marketing message. To create this message, the supplier uses his marketing concept and the manufacturer tries to improve it (MC^{SM}) by replacing one or two of the supplier's triples with triples of his own kene set that have the same K with a higher A. This implicates that the manufacturer uses his knowledge base to reduce the distance between the supplier and the applicator.

In the model, the joint marketing concept then describes the source the supplier and the manufacturer use for their marketing attempt to jointly implement the innovation. Developing this marketing concept into a joint message (*joint M*) is a mapping procedure where the *Ks* and the *As* of the joint concept are used to compute an index number that represents the joint message [see Eq. (3.14)].

$$joint\,M = \sum_{i \in MC^{SM}} (K_i \times A_i) \tag{3.14}$$

As in the case of non-cooperative VCM, the marketing concept of the applicator is also used to generate his expected message [cf. Eq. (3.3)].

Calculation of the Cooperative Marketing Cost 1 ($\mathrm{Cost1_C}$)

Similar to non-cooperative VCM, the first cost element that the supplier spends is the cost to address an applicator. The supplier cooperates with a manufacturer in order to transmit a joint message. The first cost factor in cooperative VCM is proportional to the difference of *joint M* and *EM* as depicted in Eq. (3.15). The equation is scaled using a constant ($C1_C$) to enable a comparison to other cost factors. More details on the cost calculation can be found in the Appendix.

$$Cost1_C = (C1_C \times |joint\,M - EM|) \tag{3.15}$$

Evaluation Process

As in non-cooperative VCM, the value of the innovation needs to be evaluated. However, the joint marketing concept (MC^{SM}) is used to describe the innovation to the applicator. As the manufacturer supports the supplier, the value as perceived by the applicator is contributed by both the supplier and the manufacturer as shown in Eqs. (3.16), (3.17), and (3.18). The value of the innovation is proportional to the knowledge fields (K_j) and the expertise (E_j) used in the joint marketing concept (MC^{SM}). The limitation still holds: the amount of knowledge must be at least two ($A_j \in \{2, 3\}$) so that the applicator could sense the value of the innovation.

$$CV_C = \sum_{j \in MC^{SM}} (cv_j) \tag{3.16}$$

where

$$cv_j = K_j \times E_j \quad \text{if} \quad A_j \in \{2, 3\} \quad j \in MC^{SM} \tag{3.17}$$

$$cv_j = 0 \quad \text{if} \quad A_j \in \{1\} \quad j \in MC^{SM} \tag{3.18}$$

The customer value expected by the applicator (ECV_C) can be explained in the same way as in the case of non-cooperative VCM. The way it is calculated is also similar [cf. Eqs. (3.8), (3.9), and (3.10)].

Comparison of the Offered and the Expected Customer Value

An innovation is accepted if the value offered by the supplier and the manufacturer (CV_C) is higher than or equal to the value expected by the applicator (ECV_C). Otherwise, the innovation is rejected. This evaluation process follows Eq. (3.19). It is the less critical step in cooperative VCM as the manufacturer supports the supplier during the whole interaction process. This implies that the supplier can profit from the manufacturer's experience in interacting with the applicator.

$$Acceptance : \begin{cases} yes, if\ CV_C \geq ECV_C \\ no, if\ CV_C < ECV_C \end{cases} \tag{3.19}$$

Once again, at this stage, the supplier has performed a full VCM process and has interacted with other agents in the value chain. Therefore, the supplier's kene set is improved to reflect this learning process. The amount of knowledge and expertise of the triples used in the joint marketing concept are increased by one.

Calculation of the Cooperative Marketing Cost 2 ($\mathrm{Cost2}_C$)

The second cost factor in cooperative VCM represents the cost needed to explain the value of the innovation to the applicator. Similar to the second cost factor in non-cooperative VCM, the cost is proportional to the difference between CV_C and ECV_C. The calculation of this cost factor follows Eq. (3.20). This equation is similar to other cost factor equations. The usage of the constant ($C2_C$) follows the same logic. More details on the cost calculation can be found in the Appendix.

$$Cost2_C = (C2_C \times |CV_C - ECV_C|) \tag{3.20}$$

3.4.5 Input and Output

Inputs are the environmental conditions that influence the output of the simulation. In the VCM model, there are several inputs whose values can be set by users. These inputs are listed in Table 3.5.

The marketing strategy (m) is selected by users to simulate cooperative VCM, non-cooperative VCM, or a mixture of both marketing strategies. The number of suppliers (N_S), the number of manufacturers (N_M), and the number of applicators (N_A) can be adjusted to build different environments with different proportions of agent types. Some suppliers can be set to have more capital than the rest of the suppliers using the big supplier's ratio (b) input. This input is needed to create a more similar situation to the real world where there are some suppliers that are bigger than others. In addition, there is an innovation rate (r) input where the user can set the number of supplier innovations or the number of active suppliers in the

Table 3.5 List of inputs

Symbol	Description	Value
m	VCM strategy	(Coop., non-coop., random)
N_S	Number of suppliers	$0 \leq N_S \leq 100$
N_M	Number of manufacturers	$0 \leq N_M \leq 100$
N_A	Number of applicators	$0 \leq N_A \leq 100$
b	Big supplier's ratio	$0 \leq b \leq 50$ %
r	Innovation rate (number of active suppliers)	$0 \leq r \leq 50$ %
c	Marketing capital	$0 \leq c \leq 50$ %

model. A supplier is active if he has an innovation and tries to promote it. At last, users can set the initial marketing capital (c) owned by the suppliers.

The outputs are the outcomes that are obtained after running the simulation model. In order to achieve the objective of this model, there are a number of outputs that need to be analyzed as listed in Table 3.6. The marketing success (S) is the result of the marketing attempt whether the innovation is accepted and implemented or rejected. The implementation time (IT) describes the total number of attempts by the supplier to implement his present innovation. It thus describes the duration of the implementation process. Finally, the marketing costs are to be recorded as the spent as total implementation costs (IC).

3.4.6 Multi-agent Features

The main process in the model has been described. It serves as a generic VCM process. In the real business environment, there is always more complexity. One aspect that should be included in the model is the competition between agents. To reproduce the competition aspect, some additional features are added. These features basically enable the agents to adapt and learn during the marketing process. These will also bring more complexity in the agent's interactions.

The first feature is the supplier's preference in selecting a (potential) business partner. Once a supplier is successful in addressing an agent (i.e. manufacturer or applicator), he will remember that and will try to target this agent again. This procedure is based on the *SKIN model* where previously good experience with former contacts militates in favor of renewing a partnership (see Ahrweiler et al. 2004). Schulze (2012) also suggests that a firm should select a partner that it is familiar with in order to get optimal results in collaboration.

Applicators also have their preference. It is possible that more than one supplier target an applicator and try to promote their innovations. In these cases, the applicator would definitely like to listen to the offers of the different suppliers and try to find which one is most suitable to his needs. Therefore, applicators are given the ability to listen and speak to a maximum of three suppliers. Each supplier goes through the same VCM process, but the final selection is based on the

Table 3.6 List of outputs

Symbol	Description	Value
S	Marketing success	(Acceptance, rejection)
IT	Implementation time	$n \in Z$
IC	Implementation costs	$Sum \in R$

customer value offered by the supplier. In fact, the supplier innovation that offers the highest customer value will be accepted by the applicator. Regarding the *SKIN model*, the user selects the product with the cheapest price (see Ahrweiler et al. 2004).

3.5 Conclusions and Outlook

The focus of this paper is to provide a theoretical and conceptual foundation for the model of VCM. Besides our pilot study and our multiple case studies, we rely on ABM as a methodology grounded on micro interactions among heterogeneous value-chain actors. ABM offers an additional level of realism and gives us the ability to triangulate our findings. For this purpose, we describe the concept of our agent-based model in detail by focusing on the individual decision processes. The model aims to improve the understanding of the complex processes by explaining the emergence of macro dynamics deriving from micro relations. It goes beyond previous attempts to explain the marketing of innovation and advances from former models as it uses empirical calibrated kenes and decision rules. The decision to accept or reject a supplier innovation depends on an evaluation process with corresponding fitness functions—the function of marketing objectives, messages, and customer values.

Performing experiments on the simulation offers new insights to evaluate the influence of the newness of supplier innovation, the knowledge of the agents, the initial distances or overlaps in their kenes, the adaptation of the kenes on the marketing output, as well as the interactions of the agents. The next steps in the research framework include the implementation and verification of the VCM model as well as the analysis and interpretation of the simulation results.

Appendix

Calculation of the Non-cooperative Marketing Cost 1 ($\mathrm{Cost1}_{\mathrm{NC}}$)

- Max. $Cost1_{NC} = 15$
- Condition: $K \geq 3$
- Supplier's message: $K_1^S \times A_1^S + K_2^S \times A_2^S$ (min: $3 \times 1 + 3 \times 1 = 6$; max: $6 \times 3 + 6 \times 3 = 36$)

- Applicator's message: $K_1^A \times A_1^A + K_2^A \times A_2^A$ (min: $3 \times 1 + 3 \times 1 = 6$; max: $6 \times 3 + 6 \times 3 = 36$)
- $\Delta M_{min} = 0$, $\Delta M_{max} = 30 \rightarrow C1_{NC} = 15/30$

Calculation of the Non-cooperative Marketing Cost 2 ($Cost2_{NC}$)

- Max. $Cost2_{NC} = 15$
- Condition: $K \geq 3$ *and* $A \geq 2$
- Offered customer value: $K_1^S \times E_1^S + K_2^S \times E_2^S$ (min: 0 if $A < 2$; max: $6 \times 3 + 6 \times 3 = 36$)
- Expected value: $K_1^A \times E_1^A + K_2^A \times E_2^A$ (min: 0 *if* $A < 2$; max: $6 \times 3 + 6 \times 3 = 36$)
- $\Delta CV_{min} = 0$, $\Delta CV_{max} = 36 \rightarrow C2_{NC} = 15/36$

References

Ahrweiler P, Pyka A, Gilbert GN (2004) Simulating knowledge dynamics in innovation networks (SKIN). Volkswirtschaftliche Diskussionsreihe 267, Institute of Economics, University of Augsburg

Ahrweiler P, Pyka A, Gilbert GN (2011) A new model for university links in knowledge-based economies. J Prod Innovat Manag 28(2):218–235

Asheim B (2007) Differentiated knowledge bases and varieties of regional innovation systems. Innovat Eur J Soc Sci Res 20(3):223–241

Cohen WM, Levinthal DA (1989) Innovation and learning: the two faces of R&D. Econ J 99(397): 569–596

Conti A, Hoisl K (2012) The speedy road to success: knowledge overlap in R&D teams. In: Proceedings of the DRUID 2012, Copenhagen

Cowan R, Jonard N (2009) Knowledge portfolios and the organization of innovation networks. Acad Manag Rev 34(2):320–342

Delre SA, Jager W, Janssen MA (2007) Diffusion dynamics in small world networks with heterogeneous consumers. Comput Math Organ Theor 13(2):185–202

Dosi G, Faillo M, Marengo L (2003) Organizational capabilities, patterns of knowledge accumulation and governance structures in business firms: an introduction. Laboratory of Economics and Management, Sant'Anna School of Advanced Studies, Pisa

Duncan T, Moriarty SE (1998) A communication-based marketing model for managing relationships. J Market 56(2):1–13

Fern EF, Brown JR (1984) The industrial/consumer marketing dichotomy: a case of insufficient justification. J Market 48(2):68–77

Flint DJ, Woodruff RB, Gardial SF (1997) Customer value change in industrial marketing relationships. Ind Market Manag 26(2):163–175

Fussell SR, Krauss RM (1989) Understanding friends and strangers: the effect of audience design on message comprehension. Eur J Soc Psychol 19(6):509–525

Garcia R (2005) Uses of agent-based modeling in innovation/new product development research. J Prod Innovat Manag 22(5):380–398

Garcia R, Atkin T (2005) Coopetition for the diffusion of resistant innovations: a case study in the global wine industry using an agent-based model. Working paper 05-002, Northeastern University, Institute for Global Innovation Management

Garcia R, Jager W (2011) Agent-based modeling of innovation diffusion. J Prod Innovat Manag 28(2):148–151

Gilbert GN, Pyka A, Ahrweiler P (2001) Innovation networks – a simulation approach. J Artif Soc Soc Simulat 4(3):8

Gobeli DH, Brown DJ (1987) Analyzing product innovations. Res Manag 30(4):25–31

Grimm V, Berger U, Bastiansen F, Eliassen S, Ginot V, Giske J et al (2006) A standard protocol for describing individual-based and agent-based models. Ecol Model 198(1–2):115–126

Grimm V, Berger U, DeAngelis DL, Polhill JG, Giske J, Railsback SF (2010) The ODD protocol: a review and first update. Ecol Model 221(23):2760–2768

Grunert KG, Fruensgaard Jeppesen L, Jespersen KR, Sonne A-M, Hansen K, Trondsen T, Young J (2005) Market orientation of value chains: a conceptual framework based on four case studies from the food industry. Eur J Market 39(5/6):428–455

Hillebrand B, Biemans WG (2005) Strategies for dealing with derived demand. Working paper series on research in relationship management, Nijmegen School of Management

Hillebrand B, Biemans WG (2011) Dealing with downstream customers: an exploratory study. J Bus Ind Market 26(2):72–80

Janssen MA, Ostrom E (2006) Empirically based, agent-based models. Ecol Soc 11(2):37

Jeannet J-P (2006) Value chain marketing. In: Kashani K (ed) Beyond traditional marketing: innovations in marketing practice. Wiley, Chichester, pp 19–59

Kashani K (2006) Beyond traditional marketing: innovations in marketing practice. Wiley, Chichester

Kleinaltenkamp M, Frauendorf J, Kähm E (2007) Business-to-business markets – status quo and future trends. J Bus Market Manag 1(1):7–39

Kleinschmidt EJ, Cooper RG (1991) The impact of product innovativeness on performance. J Prod Innovat Manag 8(4):240–251

Narver JC, Slater SF (1990) The effect of a market orientation on business profitability. J Market 54(4):20–35

North MJ, Macal CM, St. Aubin J, Thimmapuram P, Bragen M, Hahn J et al (2010) Multi-scale agent-based consumer market modeling. Complexity 15(5):37–47

Prabhu JC, Chandy RK, Ellis ME (2005) The impact of acquisitions on innovation: poison pill, placebo, or tonic? J Market 69(1):181–193

Ram S, Sheth JN (1989) Consumer resistance to innovations: the marketing problem and its solutions. J Consum Market 6(2):5–14

Rand W, Rust RT (2011) Agent-based modeling in marketing: guidelines for rigor. Int J Res Market 28(3):181–193

Rogers EM, Bhowmik DK (1971) Homophily-heterophily: relational concepts for communication research. Public Opin Q 34(4):523–538

Rogers EM, Kincaid DL (1981) Communication networks: towards a new paradigm for research. Free Press, New York

Schramm ME, Trainor KJ, Shanker M, Hu MY (2010) An agent-based diffusion model with consumer and brand agents. Decis Support Syst 50(1):234–242

Schulze A (2012) Distance between partner's knowledge base components as an explanation of joint innovation performance. In: Proceedings of the DRUID 2012, Copenhagen

Sheth JN, Sisodia RS (2002) Marketing productivity issues and analysis. J Bus Res 55(2):349–362

Veryzer RW (1998) Discontinuous innovation and the new product development process. J Prod Innovat Manag 15(4):304–321

Ziamou PL (1999) The effect of the degree of newness of a "really new" product on consumers' judgement. Adv Consum Res 26(1):368–371

Chapter 4
Micro Strategies and Macro Patterns in the Evolution of Innovation Networks: An Agent-Based Simulation Approach

Matthias Müller, Tobias Buchmann, and Muhamed Kudic

Abstract We focus on the evolution of interfirm innovation networks which are composed of and driven by heterogeneous actors that follow a number of well-defined cooperation partner selection strategies. For this purpose, we simulate micro level firm behaviour which shapes the macro level network patterns. In order to analyse the micro-macro link we build an agent-based simulation model (ABSM) which allows us to test causal relationships between firm strategies and the emerging network structures. With this model we analyse the structural consequences of homophily, reputation and cohesion mechanisms in a situation of information scarcity. We start with a simple model which is in a next step extended by adding a market mechanism which links the knowledge base of a firm with the reward a firm receives and with the incentives to cooperate. We show that (a) a transitive closure mechanism combined with a tendency for preferential attachment produces networks that show both, small world characteristics as well as a power law degree distribution; (b) diversity in the selection of cooperation partners is an important determinant of innovative performance if we consider an evolving network.

M. Müller (✉) • T. Buchmann
University of Hohenheim, 70593 Stuttgart, Germany
e-mail: m_mueller@uni-hohenheim.de; tobias.buchmann@uni-hohenheim.de

M. Kudic
Department Structural Economics, Halle Institute for Economic Research, Kleine Maerkerstrasse 8, 06108 Halle (Saale), Germany

Statistical Department, Stifterverband, Rellinghauser Str. 3, 45128 Essen, Germany
e-mail: muhamed.kudic@stifterverband.de

N. Gilbert et al. (eds.), *Simulating Knowledge Dynamics in Innovation Networks*, Understanding Complex Systems, DOI 10.1007/978-3-662-43508-3_4,

4.1 Introduction

Meanwhile, innovation processes are frequently studied through the lenses of network analysis techniques. Networks allow taking explicitly into account the dependency between actors driving innovation processes. In innovation networks the interaction of collaborating actors spurring knowledge exchange and mutual learning is regarded as a key driver for innovative success (Pyka 1999; Malerba and Vonortas 2009). It is an obvious prerequisite for any form of collaboration to initially find an appropriate partner. Empirical studies provide sound evidence that firms follow a broad variety of strategies in this partner search procedure. For instance, they try to connect to an actor which has the highest level of reputation and presumably the most valuable knowledge a partner can gain access to. Or, they try to find a partner that is similar in some key characteristics (e.g. in its structural embeddedness) which makes a common understanding of problems and solutions more likely. Our results connect the micro and macro level and show how the individual behaviour on the micro level shapes the evolution of structure on the macro level and how it is responsible for particular network patterns.

Network topologies affect the exchange of information, ideas and knowledge. For example, researchers have demonstrated that networks with comparably short path lengths and high clustering—so called small world networks—have a significant impact on the innovativeness of the involved firms (Schilling and Phelps 2007). Our study contributes to a deeper understanding of systemic determinants of firm innovativeness by explaining the micro level drivers of macro level properties of innovation networks.

The agent-based simulation model can be considered as a tool to disentangle the relationship between firm strategy and the emerging network. In particular, we analyse which network structure emerges if we employ a certain partner selection strategy. By adding a market mechanism we are able to simulate adaptive actor behaviour stimulated by market rewards. The basic ingredient for the development of the market mechanisms is the so called KENE approach (Gilbert 1997; Gilbert et al. 2001, 2007; Pyka et al. 2007) which we apply to model a firm's knowledge base, learning processes and the transfer of knowledge in network structures.

With our modelling exercises we seek to address three fields: We are (1) curious to understand which macro structures emerge from simple micro strategies by investigating the mechanisms of tie formation. Referring to the work of Barabasi and Albert (1999) and Watts and Strogatz (1998) we were particularly interested whether there is a simple but realistic cooperation routine which combines the advantages of both, small world and scale-free properties. For this purpose, we repeatedly analyse three important parameters characterising a network: First, the clustering coefficient informs about the cliquishness of a network; it is the tendency to build cohesive subgroups. Second, the shortest average path length is an indicator for the speed of knowledge transfer in the network.[1] Third, the degree distribution gives the probability distribution for the number of links an actor has for the entire network. Furthermore, we extend the base model by adding a market mechanism layer which generates rewards for firms dependent on the product type and product quality they manage to produce. This enables us to (2) analyse the influences of market conditions on micro strategies and the way they shape the network structures; and it allows us to (3) compare different firm strategies with regard to their (simulated) market performance. An important result of the simulation study is that a broader variety of cooperation partners yields higher market rewards.

The chapter is divided as follows: In Sect. 4.2 we present theoretical aspects which illustrate the general approach of our model, our assumptions and the simulated firm strategies. In Sect. 4.3, we describe the actual model and analyse and interpret the results. Finally, we conclude with a number of remarks and an outline of our future research agenda.

4.2 A Knowledge-Based Approach to Evolutionary Network Change

An important contribution of Marshall (1920) to the theory of the firm is the observation that each firm possesses a unique set of relations with other firms or research organisations which makes it difficult to define the boundaries of firms. These relations spur innovation processes by providing access to the knowledge base of other organisations, thereby constituting an element of a firm's organisational capital. In a similar vein, Gulati (2007) argues that a firm's web of collaborative relationships has to be considered as a firm-specific strategic resource. For this reason, network embeddedness is a vital asset for firms constituting a firm-specific element of heterogeneity (Loasby 2001; Granovetter 1985).

4.2.1 The Collective Nature of Innovation

We suggest for the design of our model that a firm's cooperative behavior is driven by the search for new and potentially valuable knowledge. Early proponents of this concept are Marshall (1920) who recognizes knowledge as the decisive factor in production processes and Penrose (1959) who identifies the knowledge base of a firm as a key resource. In the early 1980s this approach is taken up by Neo-Schumpeterian scholars (Pyka and Fagiolo 2005; Hanusch and Pyka 2007; Hodgson et al. 1994; Dopfer 2005). Here, the role of knowledge for economic

[1] Small world networks are characterized by a small average shortest path length and a relatively high level of clustering (Watts and Strogatz 1998).

development and the success of firms is explicitly recognized and constitutes the cornerstone of economic analysis. In the Neo-Schumpeterian perspective, networks are seen as a central determinant in the industrial creation of novelty, and are considered as a decisive coordination mechanism. In networks new technological opportunities are created via technological complementarities and synergies, bringing together a wide spectrum of technological and managerial competencies. Knowledge is considered no longer as a pure public good but as local, tacit, firm-specific and complex (Dosi 1988). These characteristics hamper technological knowledge from being easily exchanged on markets. Technological spillovers are no longer freely available as in the standard models of industrial organization but have to be acquired actively by participating in innovation networks (Geroski 1995). In other words, technological spillovers are spurred by innovation network embeddedness.

Researchers in economics and management science devote considerable efforts to investigate the causes, motivations and advantages for the emergence of interfirm networks and strategic alliances (e.g. Ahuja 2000; Hagedoorn 2002; Powell et al. 1996). A large number of studies focus on motives of cooperative behaviour; or they relate structural characteristics of the network, network subgroups or single actors (e.g. centrality) to outcome measures like innovativeness or performance (Gulati 1998). However, studies linking the cooperation partner selection strategies with the evolution of a network are less common. Only a small number of empirical studies have analyzed network formation and evolution (Walker et al. 1997; Orsenigo et al. 2001; Gay and Dousset 2005; Powell et al. 2005). Longitudinal empirical network studies have yet some notable limitations (for an exhaustive critique see: Kenis and Oerlemans 2008). For instance, firms tend to report pro-actively tie formations whereas tie terminations are commonly suppressed. The agent-based approach allows compensating for some of the most salient limitations of empirical research settings. Consequently, in this chapter we apply an agent-based simulation model to explicitly study the relationship between firm strategies on a micro level and network evolution on the aggregate level.

4.2.2 Modelling Adaptive Firm Behaviour

Starting with Schumpeters's "Theorie der wirtschaftlichen Entwicklung", the literature offers plenty of theoretical considerations and empirical evidence for the hypotheses that innovation and technological progress are the main drivers for economic development and growth of firms, regions and countries (e.g. Malerba 2002; Aghion and Griffith 2005). Orthodox neoclassical models provide only limited explanations that can help us to understand the relationship between cooperative behaviour, networks structures, innovation and economic performance. This is related to the oversimplified assumption of representative agents striving for profit maximization, to the illusion of a well-defined choice set to achieve this goal and to an overly abstract modelling approach which does not shed light on the core features of the innovation process as explained in this chapter. We refer to the idea of evolutionary processes in economic development characterized by uncertainty (Knight 1921), dynamic change processes instead of stable equilibria, heterogeneity of actors, bounded-rational behaviour and collective innovation processes (Pyka 1999). This evolutionary view adds a layer of considerable complexity as the emerging model is characterised by interaction and adaptive behaviour of actors, and the non-existence of stable equilibria (Nelson and Winter 1982; Holland 1996). Moreover, it paves the way to consider empirically observed processes of collective innovation in which firms and other innovative actors learn from each other and exchange knowledge (Bloch and Finch 2009).

Time has to be considered as a conditio-sine-qua-non in a model that reflects strategic behaviour and adaptation of routinized behaviour. In network structures economic actors adapt their behaviour following an internal model e.g. focussing on their R&D success (Holland 1996) in period t and subsequently change their network ties which brings the network in a different state in the following period $t+1$ (Nelson and Winter 1982). The dynamics of the emerging phenomena from this kind of adaptive behaviour are not sufficiently studied in so far. If at all, a simplified analysis is applied focusing only on patterns in a state of "behavioural" equilibrium, i.e. static patterns that are the aggregate of micro level behaviour without considering reactions of agents to a certain network configuration (Arthur 1999). In our model, we follow a different approach and explicitly consider reactions of actors to changes in the internal conditions as well as changes in the knowledge space and network configuration.

In a nutshell, we model firms as "repositories of productive knowledge" (Winter 1988, p. 169) that are able to adapt to internal and external constrains, learn from each other through interaction and seek to increase their profits. Our key interest is to analyse aggregate structures that emerge from adaptive firm behaviour.

4.2.3 Mechanisms of Network Evolution

In this study we analyse why and how different cooperation selection strategies influence the evolution of innovation networks. Following the idea of imperfect and scarce information about the value and reliability of a potential cooperation partner, we introduce a number of search strategies that aim to compensate the information deficit problem. The selection of an appropriate partner is found to be a key for the success of cooperation (Gerybadze 2004) which in turn requires a thorough scrutiny of all candidates, a procedure which is costly. Based on theoretical considerations, we suggest that the trade-off between the need for reliable information and the cost of the search process is reflected in a two-stage selection process in which firms randomly or based on the transitive closure principle pre-select a group of firms from which they make their final choice:

First, the transitive closure mechanism in network evolution leads to persistent, dense and cohesive network structures. A basic understanding of transitivity and triads (Wasserman and Faust 1994) is required to understand how the transitive closure mechanism works. A triadic subgraph has a much higher complexity level than a simple dyad. This becomes apparent when looking at the triadic states and isomorphism classes. For a triad T_{ijk}, 64 states and 16 unique isomorphism classes can be identified whereas the accurate description of these classes requires a labeling scheme specifically developed for this purpose (Wasserman and Faust 1994). Holland and Leinhardt (1971, 1977) proposed the so-called "M-A-N labeling scheme" which assigns four characters to each triad.[2] In an innovation network context, the transitive closure mechanism works as follows (Holland and Leinhardt 1971; Davis 1970): When firms operate in an uncertain environment they may use existing connections in order to gain information about potential partners. For instance, if actor *i* cooperates with actor *j* and *j* also cooperates with a third actor *k*, then *i* may get information about the trustworthiness, reliability and value of the knowledge base of *k* from *j*. Whereas, it is much more difficult and costly to collect information about other firms that are more than two steps away or not connected at all. Consequently, the probability that *i* connects to *k* is higher than the probability to connect to more distant actors. Additionally, in triadic structures the cooperative behaviour can be easily monitored as social control is strong. This prevents actors from opportunistic behaviour and thus eases the emergence of trust which is a prerequisite for the exchange of tacit knowledge (Uzzi 1997). Thus, network cohesion fosters innovation output as it promotes trust, collaboration and knowledge sharing.

Second, we apply (structural) homophily concept between firms and preferential attachment as strategies for the final selection (step two) of a cooperation partner. We also tested a pure random strategy as a benchmark, according to which firms select their partners in an arbitrary way. Homophily refers to a preference for similarity between potential cooperation partners (McPherson et al. 2001). We assume two actors to be similar if they possess a similar structural position in a network. When a firm is actively searching for a cooperation partner it may try to connect to a partner holding a central position in the network because it expects to benefit from the visibility and attractiveness that can be inferred from a central position, and to better gain itself access to the valuable knowledge of the network through the partner (e.g. the Universal Card alliance between AT&T and TSYS (Sankar et al. 1995). However, the central player might be rather reluctant to accept a peripheral actor as a partner since a less central actor will most likely benefit more from the cooperation than the more central actor. Peripheral actors might be perceived as less sophisticated concerning their knowledge base and thus are less attractive as a partner with whom one would share knowledge. Consequently, highly attractive central actors prefer to connect to other central actors rather than

---

[2] According to this scheme, the first character specifies the number of mutual dyads, the second character gives the number of asymmetric dyads, the third character displays the number of null-dyads, and the last character gives a further characterization of how the ties are directed to each other within these specific isomorphism classes by using the characters "D" (for down), "U" (for up), "T" (for transitive) and "C" (for cyclic). For further details, see Wasserman and Faust (1994).

to peripheral actors. Podolny (1994) found that the preference for actors possessing a similar status is particularly prevalent in highly uncertain environments.

Preferential attachment based on reputation is a factor that contributes more to increasing imbalances than to equalising structural positions. Reputation works as a social signal for firms searching for a cooperation partner and helps them to select a potentially valuable partner. In particular, when the level of uncertainty is high and knowledge about other firms is scarce, social signals may play a role for reducing the circle of potential partners (Lazega et al. 2012). Reputation is linked to the position in the network, i.e. we assume that the most central actor in the network possesses the highest reputation. The most central actor is well embedded in the network and has many direct and indirect ties. This guarantees that information about this actor can easily spread in the network. Additionally, Powell et al. (1996, 2005) showed that it is not success or size that brings an organisation in a central network position, but the openness and attractiveness as cooperation partner which contributes to visibility. An additional element, namely costs which accrue for searching and evaluating information about a potential partner, makes a selection process led by social criteria advantageous (Gould 2002). The search for the ideal cooperation partner can be cumbersome, time consuming, cost intensive and requires capabilities of judgment. Typically small firms lack the necessary skills and resources to conduct extensive search procedures that would cover all potential partners and produce reliable information for the evaluation (Giuliani 2010). We operationalize this reputation effect by a preferential attachment mechanism leading to a "rich get richer" kind of degree distribution in the network. That is, firms which have the largest number of ties in the network are preferably chosen as partners (Barabasi and Albert 1999; Price 1976). Barabasi and Albert (1999, p. 510) suggest that "large networks self-organize into a scale-free state" due to the growth of the network in tandem with the preferential attachment mechanism. Accordingly, the probability P(k) that a network actor cooperates with k other actors decreases according to a power law which can be expressed as $P(k) \sim k^{-\gamma}$. Different from this approach, our model focuses directly on the actors and their strategic behaviour and less on connecting probabilities as such. Thus, it is more of a bottom-up perspective based on the individual behaviour and less focused on holistic aspects. With regard to the analysis of the degree distribution we want to elaborate if one of our strategies leads to a power law degree distribution which has been observed in empirical studies, e.g. in the biotechnology industry (cf. Powell et al. 1996, 2005).

4.3 An Agent-Based Simulation Approach

The following section describes the model and explains some of our simulation results.

4.3.1 *The Simulation Model*

For the simulation of the evolution of interfirm innovation networks we follow a two-stage approach. First, we construct a simple model capturing a variety of cooperation partner selection strategies. These strategies are determined by cohesion, homophily and reputation considerations. By doing so, we intend to implement realistic but simple cooperation strategies that are frequently cited in strategic management, sociology and economics literature. Second, we increase the complexity of our model by adding a market mechanism which influences the decisions of firms whether to cooperate or solely rely on internal R&D activities. In this environment, firms produce and sell their products based on the success in finding new knowledge through R&D and cooperation.

4.3.1.1 The Selection of Cooperation Partners

A key element of our modelling approach is the assumption of general information scarcity concerning the value and reliability of potential partners. Considering the high efforts and costs for gaining and evaluating very detailed information about possible partners (their products, technologies, complementary resources etc.) we assume that firms select their partner in a two stage process. With a first routine, a firm builds a pool of pre-selected potential partners that are either arbitrarily chosen or chosen according to the transitive closure principle. This reflects the limited information gaining and processing capacity of a firm but also the accruing transaction costs. For the case of the transitive closure pre-selection routine, we assume firms to choose among organisations which they know already through existing information channels, i.e. they chose among firms which are already connected indirectly through a partner.

Through a second routine, a more intense analysis of the firms being part of the pre-selection pool takes place and results in the final partner selection. Here we distinguish between a preferential attachment, a homophily strategy and a situation of pure random selection. The preferential attachment strategy implies a preference for the firm with the highest number of already existing ties. Whereas, with the homophily strategy a firm seeks partners with a rather similar number of cooperation partners.

With the two stages and the different strategies on each stage we can define the six experiments illustrated in Table 4.1.

4.3.1.2 Production of Goods

Inspired by the work of Gilbert et al. (2001, 2007) and Pyka et al. (2007) we also implemented a market mechanism. To embed a dynamic market environment, we start with a model design which allows for the production of goods based on the

Table 4.1 Pre-selection and selection strategies

No.	Pre-selection	Selection
1	Random	Preferential attachment
2	Random	Homophily
3	Random	Random
4	Transitive closure	Preferential attachment
5	Transitive closure	Homophily
6	Transitive closure	Random

individual knowledge base of a firm. Key characteristic of this approach is the mapping of knowledge in form of a so-called "*kene*" (Gilbert 1997). Drawing on the idea of genes in biology, the *kene*-concept represents the entire knowledge base K of a firm i by a number of single units of knowledge [Eq. (4.1)]. Each knowledge unit is described by the information about the general technological field (Capability C), the particular technological trajectory followed in that field (Ability A) and a corresponding expertise level (Expertise E) (cf. Gilbert et al. 2001).

$$K_i = \begin{pmatrix} C_1 \\ A_1 \\ E_1 \end{pmatrix}, \begin{pmatrix} C_2 \\ A_2 \\ E_2 \end{pmatrix}, \begin{pmatrix} C_3 \\ A_3 \\ E_3 \end{pmatrix}, \ldots, \begin{pmatrix} C_n \\ A_n \\ E_n \end{pmatrix} \tag{4.1}$$

Furthermore, we assume that the knowledge owned by a firm will not be entirely used for the production of a certain good. There are pieces of knowledge in the knowledge base which are well known to the firm, but are at a particular moment in time not necessary for production [Eqs. (4.2) and (4.3)]. Knowledge pieces which are used for the production of goods are part of a so-called innovation-hypothesis (Gilbert et al. 2001; Ahrweiler et al. 2004).

$$K_1 = \begin{pmatrix} 1 \\ 4 \\ 2 \end{pmatrix}, \begin{pmatrix} \mathbf{47} \\ \mathbf{8} \\ \mathbf{9} \end{pmatrix}, \begin{pmatrix} 56 \\ 2 \\ 7 \end{pmatrix}, \begin{pmatrix} \mathbf{143} \\ \mathbf{9} \\ \mathbf{1} \end{pmatrix} \tag{4.2}$$

$$\Downarrow$$

$$IH_1 = \begin{pmatrix} 47 \\ 8 \\ 9 \end{pmatrix}, \begin{pmatrix} 143 \\ 9 \\ 1 \end{pmatrix} \tag{4.3}$$

To create a market on which firms compete with heterogeneous products, we implemented a computation for both the product group (ID) and a product quality (PQ) based on the individual knowledge of a firm [Eqs. (4.4) and (4.5)]. We calculate the product ID as the sum of the capabilities used in the innovation-hypothesis modulo the number of possible products. For example a firm with the capabilities $c_{1,2,3} = 47, 143$ will produce the product with the number 47 + 143 = 190 *modulo* 50 = **40**, if we consider 50 possible products.

$$ID = \sum C_{IH} \; modulo \; N \, Products \tag{4.4}$$

To calculate the product quality, we used the fact that one product can be produced by several combinations of knowledge units and each combination of knowledge has a natural limited maximum of the product quality level. We consider the quality PQ_i of a product to be determined by a firm's expertise, abilities and capabilities. We therefore created three equal factors (each ranging from 0 to 1), which, multiplied with a parameter *maximum level of quality (max PQ)*, define the individual product quality.

$$PQ_i = QE_i * QA_i * QC_i * \max PG \tag{4.5}$$

$$QE_i = \frac{\sum E_i}{\sum E_{max}} \tag{4.6}$$

$$QA_i = \frac{\sum A_{IH} \; modulo \; 11}{10} \tag{4.7}$$

$$QC_i = \frac{\sum C_{IH}}{\sum C_{max}} \tag{4.8}$$

The first factor *QE* refers to the concept of learning by doing and is simply the mean expertise of the firm's innovation hypothesis divided by the maximum of possible mean of expertise [Eq. (4.6)]. For the second factor *QA* we take the sum of the abilities in the innovation-hypothesis modulo 10 and then divided the result by 10 [Eq. (4.7)]. So far, both factors can be improved to the maximum only by learning by doing or incrementally altering the abilities. The concern here is that the occurring market dynamic would be very limited. The firm with the best ability combination and highest level of expertise would be the firm with the highest reward and with a high probability would stay so. To create a more dynamic market we account for the fact that one product can be produced by several combinations of knowledge units and hence we created a third factor *QC*. This factor determines the maximum level of quality a firm can actually achieve with a certain combination of knowledge and is the sum of all capabilities used in the *IH* divided by the highest possible sum [Eq. (4.8)].

As a result, the maximum product quality a firm can achieve through learning by doing or incremental innovation is limited by the used knowledge. Thus, new firms can enter an already existing market by finding a new combination of capabilities which may yield a higher maximum level of quality. In this way we expect to find a certain pattern of an increasing quality over time, divided into phases (or product generations), where each phase is represented by a certain (at that moment) leading combination of knowledge.

4.3.1.3 Market Routine and Firm Innovation Strategies

The next elementary building block of our model is a market routine which functions as an advisory institution for firms with regard to their innovation strategies. The market routine we use is a static reward system neglecting the versatile demand side of an economy. Within a certain product category the routine simply rewards firms based on their product quality decreasingly, starting with a fixed sum for the firm with the highest product quality.

As a consequence of incomplete information and bounded rationality, firms' decisions about their R&D and cooperation efforts are determined only by the reward they receive. Following an idea of routines and satisfying behaviour (Pyka 1999) we assume two reward thresholds to simulate a firm's adaptive behaviour. The first threshold—a radical threshold—determines whether a firm will search for a new product, i.e. they will try to learn new capabilities though R&D or cooperation (radical innovation) and will randomly alter their innovation hypothesis to change their product. However, if the firm's reward is above the radical threshold level but below the incremental threshold level the firm will try to innovate only incrementally by altering one of its abilities. Finally, if the reward level is above the two threshold lines a firm will stop pursuing own R&D or own cooperation efforts, i.e. it will not ask other firms for a R&D partnership but will still remain open for cooperation if being asked.

We implemented three possibilities for a firm to change its knowledge base. The first possibility can be referred to as *learning by doing, i.e.* a firm increases its expertise level by actually applying its knowledge. Vice versa, *forgetting by not doing*, expresses a decreasing expertise level of knowledge and eventually the loss of the knowledge unit if it is not used. The second possibility of changing knowledge results from own R&D efforts. We consider R&D as a stochastic process leading to new knowledge units after an arbitrary amount of time. To innovate incrementally the firm develops a new ability into a random capability. However, if the firm tries to innovate radically it develops a complete new knowledge unit which is included in the knowledge base and can be picked afterwards to alter the innovation-hypothesis. Lastly, a firm may cooperate with other firms to gain new knowledge. For cooperation we distinguish between incremental and radical cooperation. Generally, the cooperation routine follows the subsequent steps: After selecting a partner the cooperation starts for a defined number of periods (which in the end limits the number of simultaneous cooperation per firm). When the cooperation ends, new knowledge is transferred to the firm which started the cooperation. Which knowledge gets transferred, depends on the kind of cooperation. For a radical cooperation, this is a randomly chosen knowledge unit. Respectively for incremental cooperation it is a new ability in a randomly chosen unit, what in turn requires that only cooperation partners possessing a knowledge unit with a similar capability are considered in the pre-selection.

Table 4.2 Reference parameters of the baseline-model

Variable	Value	Variable	Value
Firms	100	Products	20
Possible capabilities (max)	100	Incremental threshold	8
Possible abilities	30	Radical threshold	6
Maximum expertise level	20	Radical innovation probability	40
IH length	3	Incremental innovation probability	10
Max quality	50	Number of pre-selected firms	5
Max reward	10	Possible capabilities (beginning)	30

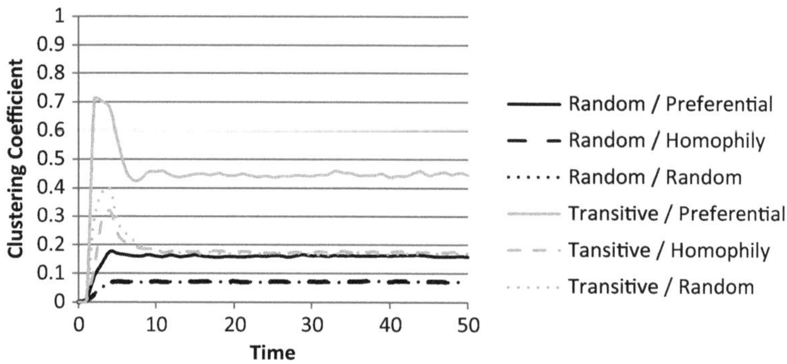

Fig. 4.1 Clustering coefficient

4.3.2 *Simulation Results and Discussion*

A simulation run starts with an initiation of the firms. The firms then apply continuously their individual cooperation rules in discrete time steps. The analytical part is based on the following steps: First, we compare the resulting network structure for different pre-selection and selection strategies without any market influences. Second, we embed the described dynamic market model and analyze the resulting changes. Each simulation is based on a specific parameter setting. For a baseline scenario we apply the values as reported in Table 4.2. Our reported simulation results are based on the average of 500 single simulation runs.

We start our analysis of the network structure with the clustering coefficient (Fig. 4.1), the shortest average path length (Fig. 4.2) and the log-log degree distribution (Fig. 4.3) for all our assumed pre-selection-selection combinations in a model without a market process. The peaks in the graphs are caused by the initialization of the simulation.

What we see is that the random-random network shows the lowest level of cliquishness. Interestingly a random-homophily strategy shows the same results although one could expect that homophily tends to create highly connected cliques. In contrast, cohesiveness through a transitive closure pre-selection increases the cliquishness slightly as one can see for the random and homophily selection. This

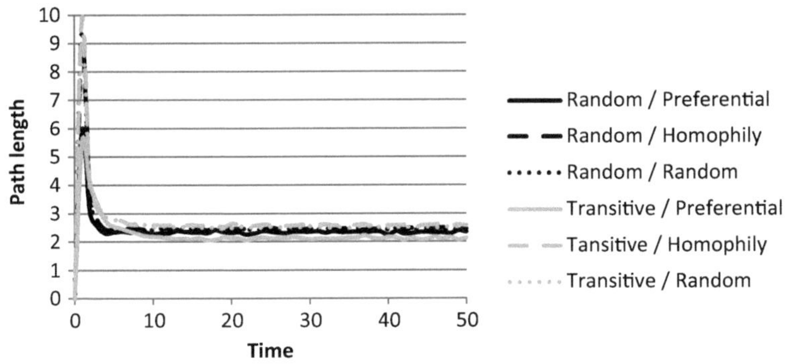

Fig. 4.2 Shortest average path length

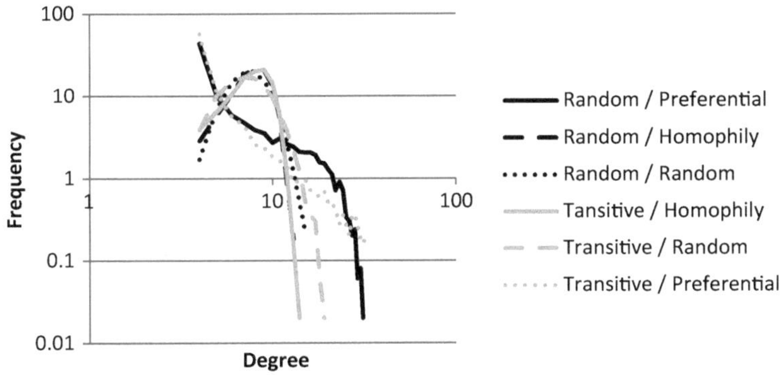

Fig. 4.3 Log-log degree distribution

shift between a random and a transitive closure pre-selection is also evident for the preferential attachment strategies. Finally we note that the preferential attachment selection itself increases the cliquishness considerably.

Considering the path length of the evolving networks (Fig. 4.2) we notify that the average path length does not differ much for all tested strategies and pre-selection situations. It is furthermore striking that the average path length is at a relatively low level pending between 2 and 3.

To investigate the degree distribution, we created a log-log graph (Fig. 4.3). Not surprisingly, all strategies without a preferential attachment effect show a Poisson distribution. Only the preferential attachment strategies (especially the transitive closure—preferential attachment strategy) show the characteristic power law distribution optimally represented by a straight line.

So far we can note that the combination of a transitive closure principle in an environment of information scarcity with a simple preferential attachment procedure shows surprising network properties. This strategy seems to produce networks which show both, small world characteristics as generated by the Watts-Strogatz

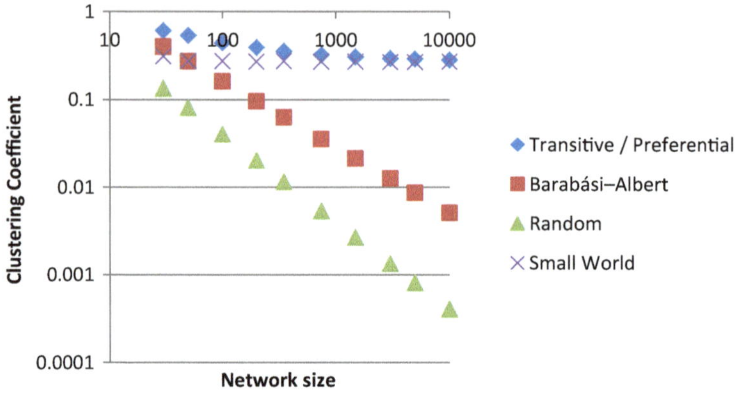

Fig. 4.4 Log-log clustering coefficient

algorithm (Watts and Strogatz 1998) as well as a power law degree distribution produced by the Barabási-Albert mechanism (Barabasi and Albert 1999). We were curious whether this holds true for other occurrences, such as the dependence of the network size on the clustering coefficient.

Although empirical results show that network size and clustering coefficient seem to by independent, the Barabási-Albert and random graphs follow a power law (Barabasi and Albert 2002, p. 31). In contrast, the results for small world networks and for our simulated networks show—especially for big networks—relative independence of the clustering coefficient from the network size (Fig. 4.4).

To analyse the behaviour and hence test the robustness of our simulation results we additionally altered systematically important variables such as the number of pre-selected firms, the network size and the maximum number of simultaneous cooperative projects. We figured out that in particular the parameters network size and number of pre-selected firms are key variables which strongly influence the emerging network structure. The subsequent figures, Figs. 4.5, 4.6 and 4.7, present the network properties for a pre-selection of 30 firms instead of 5.

For the first two plots, the qualitative results do not differ much. Again, the preferential attachment strategies are characterised by an outstanding high level of cliquishness and a short average path length. Note however that we see in Fig. 4.7 an upward slope for the right part of the two curves, which indicates that the number of firms with a medium number of cooperations decreases. The augmented number of pre-selected firms increases the chances that firms choose one of the central actors in their pre-selection, which is then chosen as partner because of the preferential attachment mechanism. As a result we presume that an increasingly pure preferential attachment strategy implemented with fewer limitations for the pre-selection (respectively information scarcity) tends to create star-networks with only few hubs and no medium connected firms.

The next block of the analysis is dedicated to the simulation results with a market environment where firms compete through innovation. With the market routine we

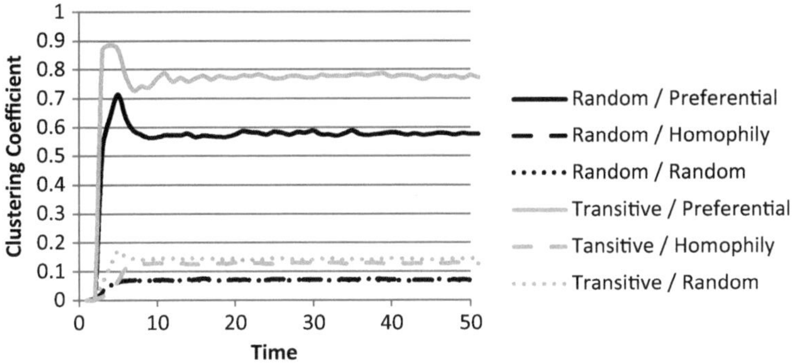

Fig. 4.5 Clustering coefficient (number of pre-selected firms = 30)

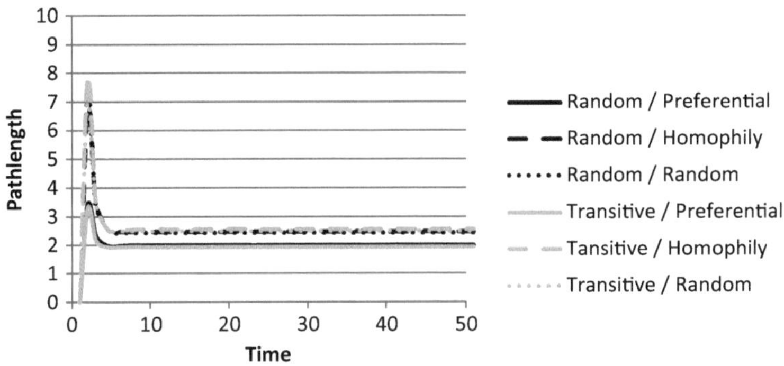

Fig. 4.6 Shortest average path length (number of pre-selected firms = 30)

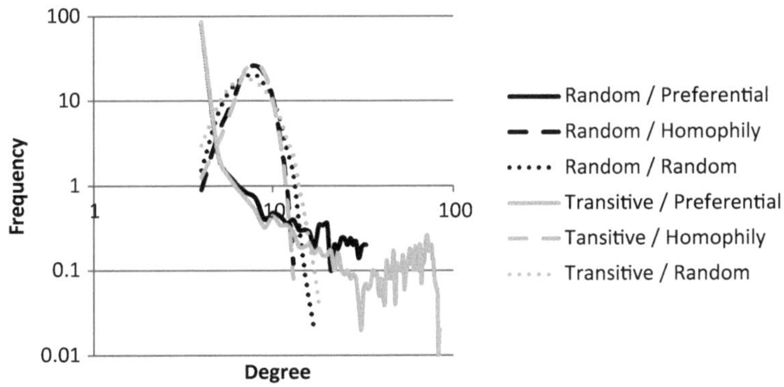

Fig. 4.7 Log-log degree distribution (number of pre-selected firms = 30)

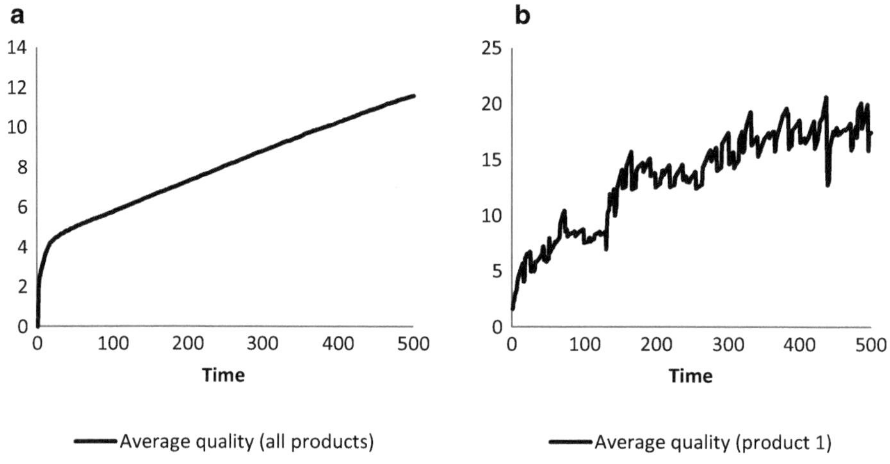

Fig. 4.8 Increasing quality levels

intended to embed the cooperation selection procedure into a more realistic and thus more complex environment where the incentives to cooperate are determined by the market success and—vice versa—the market position is a function of the success of cooperation.

As Fig. 4.8a proves, the average product quality level for all firms and products increases over time indicating that firm decisions enable them to adapt and develop their knowledge in order to achieve higher rewards. In Fig. 4.8b we present the average level of quality for the leading firms for one particular product category. Three phases of uprising quality levels are eminent, each detached by a new phase and therefore technology (i.e. knowledge combination).

If we now turn to the network structure under market conditions, we see major qualitative changes that have to be analysed carefully. First, we observe an upward shift concerning the cliquishness curve of the homophily strategy with a random pre-selection (Fig. 4.9). Second, the plotted path lengths are more dispersed and especially the path length of the random-random strategy is increased (Fig. 4.10). Finally, we see that the characteristic bell shaped curves of the degree distribution have now shifted especially at their beginning (Fig. 4.11).

At this point, it is important not to jump too rashly into first glance conclusions. We have to consider that the market creates conditions under which different strategies perform differently. The networks we created previously consist of firms who have at least four out-going ties because we assumed that every firm tries to cooperate actively. If we now implement the market routine, firms with a sufficient reward have no incentive to collaborate and hence the density of the network will decrease. Furthermore, there are now firms without any cooperation at all, which in turn decreases the size of the network.

What we see in Table 4.3 is that the size of the network decreases remarkably in dependence of the chosen strategies. This can be explained by the fact that firms

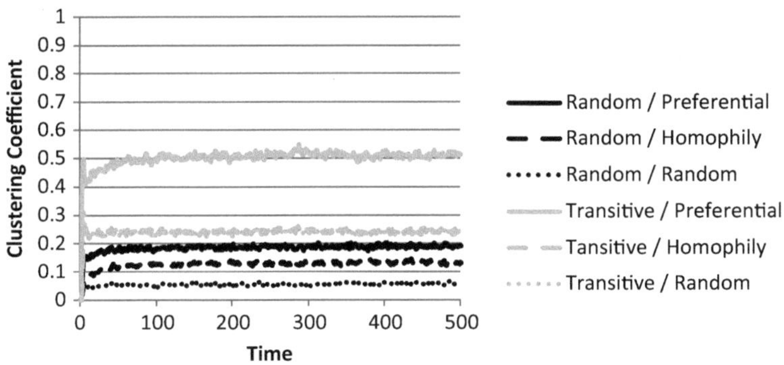

Fig. 4.9 Clustering coefficient with market (number of pre-selected firms = 5)

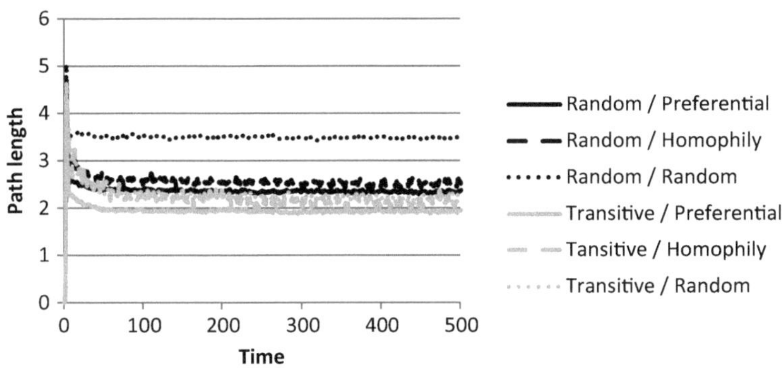

Fig. 4.10 Shortest average path length with market (number of pre-selected firms = 5)

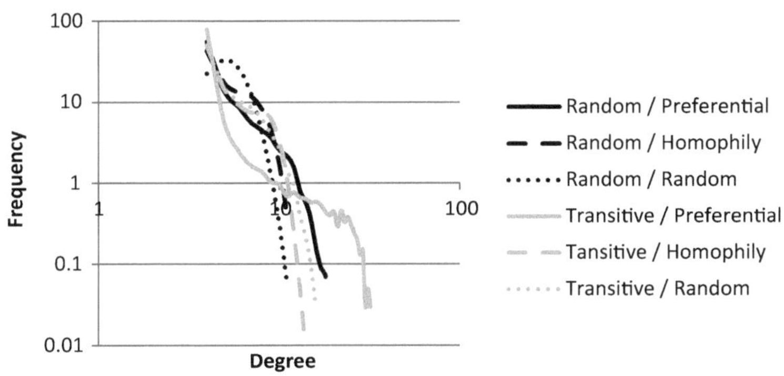

Fig. 4.11 Log-log degree distribution with market (number of pre-selected firms = 5)

which do not actively seek for a cooperation partner (because their reward is above the threshold level) must be chosen by other firms to become part of the network. If we now consider the different selection strategies, it becomes clear why some

Table 4.3 Average number of firms in cooperation

Strategy	Random/ preferential	Random/ homophily	Random/ random	Transitive/ preferential	Transitive/ homophily	Transitive/ random
Network size	47.53	55.5	87.43	41.73	47.56	47.53

strategies tend to create smaller networks than others. Firms with a preferential attachment strategy try to find a partner with a maximum number of existing ties. Consequently, firms without any cooperation will not be chosen. The homophily strategy lets firms search for firms with an equal number of ties; so again firms without any own cooperation are less likely to be chosen by the other firms. Finally, if we assume a situation of a transitive closure pre-selection in which firms without any cooperation are less likely to be pre-selected in the first place. Hence, the random strategy together with a random pre-selection is the only situation in which firms without any cooperation have a good chance to be chosen as partner.

The decreasing size of the network and the decreasing number of cooperation leads finally to changes in the overall density. In turn, the decreasing density leads to a relatively higher path length for a random-random strategy and to the changes in cliquishness. Furthermore the high number of firms without any cooperation or only passive cooperation shifts the degree distribution.

Inspired by the work of Morone and Taylor (2004) and Cowan and Jonard (2004) who created two different knowledge diffusion processes with disparate results concerning the performance of network structures, we were finally interested whether the different strategies would yield different rewards when competing directly against each other. Having in mind that in our model it is of vital importance for a firm to gain new knowledge, we expected strategies leading to short path lengths for the respective firms to be advantageous. In addition, a second factor, namely the clustering coefficient of a firm, might be relevant too. In order to ensure that the new simulation setting does not qualitatively change the observed path length and cliquishness we did another simulation run and present the occurring average node path length respectively average node cliquishness in Figs. 4.12 and 4.13 and in Fig. 4.14 the average reward of our portfolio of tested strategies.

In accordance with our expectations, we get the same qualitative results for the network structure, concerning node path length and clustering coefficient. However, the ranking order of the rewards of the strategies is unusual reverse. We expected that the strategies generating a short path length would yield the highest rewards. Surprisingly however, the strategy with the highest path length and the lowest cliquishness generates the highest rewards. Furthermore, as the ranking order indicates the reward seems negatively correlated with cliquishness and positively correlated with the path length.

Our first intention to explain the surprising results was that firms within a clique may be hampered in its technological development by a lock-in effect. Firms that share the same knowledge compete within the same product categories. In contrast, firms with a low level of cliquishness, such as firms pursuing a random–random

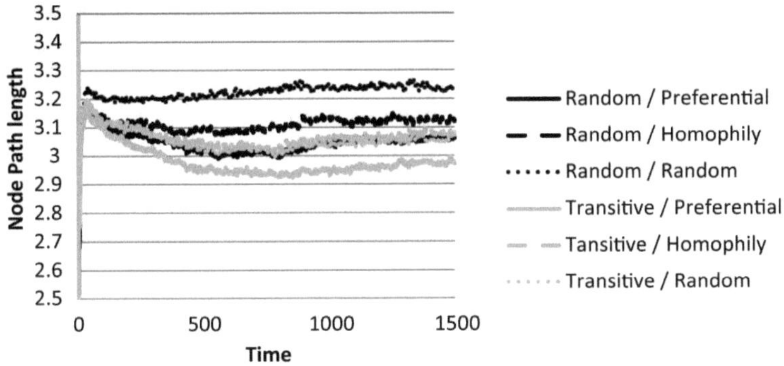

Fig. 4.12 Average node path length

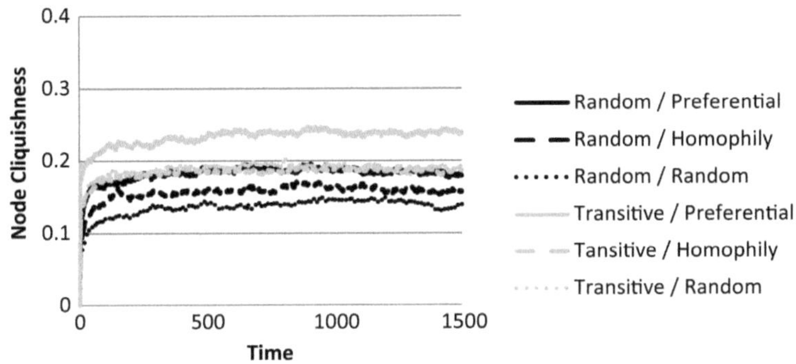

Fig. 4.13 Average node cliquishness

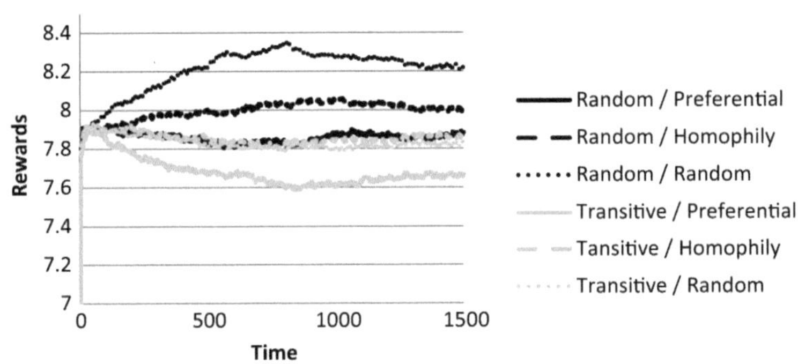

Fig. 4.14 Average reward

strategy, have access to a broader variety of knowledge thus may find a niche, without any competitors. However, against our prior expectations, a regression analysis showed that there was no significant correlation between cliquishness and reward or path length and reward. So we suppose that neither a small or high path length nor a high or low cliquishness does advantage or disadvantage the firms in our model. It turned out that it is rather a question of variety of partners that positively influences the reward level, something which can be neglected if we consider only static networks. The higher the variety of different cooperation partners, the higher is the reward the firms receive. As already stated, transitive closure, homophily and preferential attachment strategies limit the subset of potential partners because they exclude certain firms that do not possess the appropriate characteristics. Accordingly, the variety of knowledge which can be obtained is also more limited which explains the advantage of random strategies.

4.4 Concluding Remarks and Further Research Agenda

The aim of this chapter was to develop a simulation model in order to analyse the complex dynamics of innovation networks. Our starting point was the work of Watts and Strogatz (1998) and Barabasi and Albert (1999) who developed separately path breaking algorithms to generate networks with small-world characteristics respectively networks with a power law degree distribution. However, the shortcoming of both algorithms in our opinion is that on the one hand they fail to combine the network characteristics as they have been empirically observed and on the other hand their lack of a reasonable description of a firm's behaviour when it has to select a cooperation partner. Hence, we sought to find an applicable routine representing heterogeneous and individual actors which does not treat the behaviour of a firm as a game of dices. Based on the idea of a prevalent information scarcity we assumed a two stage selection process with a pre-selection and a subsequent final-selection. For the pre-selection we tested the effects of randomness as well as a transitive closure principle. For the final-selection we chose preferential attachment, homophily as well as random principles to be worth analyzing. As our result shows, if we assume a pre-selection process characterized by *transitive closure* and a following selection which implies a *preferential attachment* principle we found outstanding network characteristics, for a small number of pre-selected firms ($m = 5$). In more detail, we suggest to describe the behaviour of a firm, as follows:

Build a group of *m* possible partners. If possible, only choose partners which you know through existing links to other firms. If not, do it randomly. Finally, choose the firm with the highest degree.

In that case, the simulated network combines both small world as well as scale-free properties. Additionally, this seems to hold even for big networks or if we increase the number of links per firm. So, in contrary to Barabási and Albert's *growth* and *preferential attachment* principles, we stress the importance of *transitive closure* and *preferential attachment* under the situation of *information scarcity* to be the prevailing mechanisms when networks are created.

To capture the complex dynamics of an innovation network, we created a market procedure in a second step. It lets firms produce heterogeneous products based on their individual knowledge which they can expand by starting to cooperate. Influenced by market conditions the analysed strategies not only differ with respect to the network characteristics such as path length or clustering coefficient but also they create completely different networks with different size and density. In this respect, it became obvious that if we consider the effects of different applicable firm strategies in combination with an incentive system such as our market procedure then we have to expand the scope of analysis.

Finally we were curious to see how different strategies perform in our market procedure. Because in our model it is of vital importance to gain access to new knowledge, we expected strategies which lead to small path length or alternatively a high clustering coefficient to be advantageous. Contrary to our expectations, however, neither small path lengths nor a high clustering coefficient was advantageous. If we consider a market mechanism, such as ours, it is rather a question of variety, because strategies which lead to a broad set of different partners yield the highest rewards. So, despite the fact that small path lengths lead to fast knowledge diffusion and hence firms can have fast access to profitable knowledge, for an evolving network it may become more important to ensure a diversification of the selection of partners.

Two main avenues mark our future research agenda. First, it would be interesting to analyse the effects of the parameters such as the number of pre-selected partners, the network size and the degree. So far we can state that the results alter especially if we vary the level of information scarcity, i.e. the number of pre-selected partners. Second, we think that it is worthwhile to refine the market routine. For instance, the interplay between processes which reward cliquishness, small path length, diversification and the corresponding decision making routine of the firms should be further elaborated.

References

Aghion P, Griffith R (2005) Competition and growth: reconciling theory and evidence, Zeuthen lectures. MIT Press, Cambridge, MA

Ahrweiler P, Pyka A, Gilbert N (2004) Simulating knowledge dynamics in innovation networks (SKIN). In: Leombruni R, Richiardi M (eds) The agent-based computational approach. World Scientific Press, Singapore

Ahuja G (2000) The duality of collaboration: inducements and opportunities in the formation of interfirm linkages. Strat Manag J 21(3):317–343

Arthur WB (1999) Complexity and the economy. Science 284(5411):107–109

Barabasi AL, Albert R (1999) Emergence of scaling in random networks. Science 286:509–512

Barabasi AL, Albert R (2002) Statistical mechanics of complex networks. Rev Mod Phys 74(1): 47–97

Bloch H, Finch JH (2009) Firms and industries in evolutionary economics: lessons from Marshall, Young, Steindl and Penrose. J Evol Econ 20(1):139–162

Cowan R, Jonard N (2004) Network structure and the diffusion of knowledge. J Econ Dynam Contr 28(8):1557–1575

Davis JA (1970) Clustering and hierarchy in interpersonal relations: testing two graph theoretical models on 742 sociomatrices. Am Sociol Rev 35(5):843–851

Dopfer K (2005) The evolutionary foundations of economics. Cambridge University Press, Cambridge

Dosi G (1988) Sources, procedures, and microeconomic effects of innovation. J Econ Lit 26(3): 1120–1171

Gay B, Dousset B (2005) Innovation and network structural dynamics: study of the alliance network of a major sector of the biotechnology industry. Res Pol 34(10):1457–1475

Geroski P (1995) Markets for technology: knowledge, innovation and appropriability. In: Stoneman P (ed) Handbook of the economics of innovation and technological change. Blackwell, Oxford

Gerybadze A (2004) Technologie-und Innovationsmanagement: Strategie, Organisation und Implementierung. Vahlen, München

Gilbert N (1997) A simulation of the structure of academic science. Socio Res Online 2(2):3, http://www.socresonline.org.uk/socresonline/2/2/3.html

Gilbert N, Pyka A, Ahrweiler P (2001) Innovation networks – a simulation approach. J Artif Soc Soc Simulat 4(3):1–13

Gilbert N, Ahrweiler P, Pyka A (2007) Learning in innovation networks: some simulation experiments. Phys A 378(1):100–109

Giuliani E (2010) Network dynamics in regional clusters: the perspective of an emerging economy. Papers in Evolutionary Economic Geography (PEEG) 1014

Gould RV (2002) The origins of status hierarchies: a formal theory and empirical test. Am J Sociol 107(5):1143–1178

Granovetter M (1985) Economic action and social structure: the problem of embeddedness. Am J Sociol 91:481–510

Gulati R (1998) Alliances and networks. Strat Manag J 19(4):293–317

Gulati R (2007) Managing network resources: alliances, affiliations and other relational assets. Oxford University Press, Oxford

Hagedoorn J (2002) Inter-firm R&D partnerships: an overview of major trends and patterns since 1960. Res Pol 31(4):477–492

Hanusch H, Pyka A (2007) Elgar companion to neo-Schumpeterian economics. Edward Elgar, Cheltenham

Hodgson GM, Samuels WJ, Tool MR (1994) The Elgar companion to institutional and evolutionary economics. Edward Elgar, Aldershot

Holland JH (1996) Hidden order: how adaptation builds complexity. Basic Books, New York

Holland PW, Leinhardt S (1971) Transitivity in structural models of small groups. Comp Group Stud 2(2):107–124

Holland PW, Leinhardt S (1977) A dynamic model for social networks. J Math Sociol 1:5–20

Kenis P, Oerlemans LAG (2008) The social network perspective: understanding the structure of cooperation. In: Cropper S, Ebers M, Huxham C, Ring PS (eds) The Oxford handbook of inter-organizational relations. Oxford University Press, Oxford

Knight FH (1921) Risk, uncertainty and profit. Houghton Mifflin, New York

Lazega E, Mounier L, Snijders T, Tubaro P (2012) Norms, status and the dynamics of advice networks: a case study. Soc Networks 34(3):323–332

Loasby BJ (2001) Time, knowledge and evolutionary dynamics: why connections matter. J Evol Econ 11(4):393–412

Malerba F (2002) Sectoral systems of innovation and production. Res Pol 31(2):247–264

Malerba F, Vonortas NS (2009) Innovation networks in industry. Edward Elgar, Cheltenham

Marshall A (1920) Principles of economics: an introductory volume. Macmillan, London

McPherson M, Smith-Lovin L, Cook JM (2001) Birds of a feather: homophily in social networks. Annu Rev Sociol 27:415–444

Morone P, Taylor R (2004) Small world dynamics and the process of knowledge diffusion: the case of the metropolitan area of greater Santiago De Chile. J Artif Soc Soc Simulat 7(2):1–28
Nelson RR, Winter SG (1982) An evolutionary theory of economic change. Harvard University Press, Cambridge, MA
Orsenigo L, Pammolli F, Riccaboni M (2001) Technological change and network dynamics, Lessons from the pharmaceutical industry. Res Pol 30(3):485–508
Penrose ET (1959) The theory of the growth of the firm. Oxford University Press, Oxford
Podolny JM (1994) Market uncertainty and the social character of economic exchange. Adm Sci Q 39(3):458–483
Powell WW, Koput KW, Smith-Doerr L (1996) Interorganizational collaboration and the locus of innovation: networks of learning in biotechnology. Adm Sci Q 41(1):116–145
Powell WW, White DR, Koput KW, Owen-Smith J (2005) Network dynamics and field evolution: the growth of interorganizational collaboration in the life sciences. Am J Sociol 110(4): 1132–1205
Price DS (1976) A general theory of bibliometric and other cumulative advantage processes. J Am Soc Inf Sci 27(5):292–306
Pyka A (1999) Der kollektive Innovationsprozeß eine theoretische Analyse informeller Netzwerke und absorptiver Fähigkeiten. Duncker und Humblot, Berlin
Pyka A, Fagiolo G (2005) Agent-based modelling: a methodology for neo-Schumpeterian economics. In: Hanusch H, Pyka A (eds) The Elgar companion on neo-Schumpeterian economics. Edward Elgar, Cheltenham
Pyka A, Gilbert N, Ahrweiler P (2007) Simulating knowledge-generation and distribution processes in innovation collaborations and networks. Cybern Syst 38(7):667–693
Sankar CS, Boulton WR, Davidson NW, Snyder CA, Ussery RW (1995) Building a world-class alliance: the universal card: TSYS case. Acad Manag Exec 9(2):20–29
Schilling MA, Phelps CC (2007) Interfirm collaboration networks: the impact of large-scale network structure on firm innovation. Manag Sci 53(7):1113–1126
Schumpeter JA (1911) Theorie der wirtschaftlichen Entwicklung. Duncker und Humblot, Leipzig
Uzzi B (1997) Social structure and competition in interfirm networks: the paradox of embeddedness. Adm Sci Q 42(1):35–67
Walker G, Kogut B, Shan W (1997) Social capital, structural holes and the formation of an industry network. Organ Sci 8(2):109–125
Wasserman S, Faust K (1994) Social network analysis: methods and applications. Cambridge University Press, Cambridge
Watts DJ, Strogatz SH (1998) Collective dynamics of 'small-world' networks. Nature 393: 440–442
Winter SG (1988) On coase, competence, and the corporation. J Law Econ Organ 4(1):163–180

Part II
Testing Policy Options

Chapter 5
Simulating the Effects of Public Funding on Research in Life Sciences: Direct Research Funds *Versus* Tax Incentives

Manuela Korber and Manfred Paier

Abstract An on-going debate in the innovation policy arena revolves around the effects of public research funding. While government intervention is usually not questioned, appraising the role of direct research funds (government grants for research projects) versus tax incentives (tax exemption/deduction of research expenses) remains a core issue. In this chapter we make methodological contributions to ex-ante evaluation of these alternative government research funding instruments. Building on the SKIN model, we develop an agent-based simulation of a localized life sciences innovation system (Vienna, Austria). Companies, universities, public research and other relevant research organizations are modelled as heterogeneous agents that make investment decisions about conducting research, exchange assets with other agents and produce knowledge output. Simulation runs refer to a 30 year period, distinguishing three funding scenarios: *Direct funding (no tax incentives)*, *tax incentives (no direct funding)* and the *co-occurrence of both (direct funding and tax incentives)*. First simulation results for the Vienna life sciences innovation system suggest that the overall volume of required public funds could be lower for tax incentives than for direct funding. However, we find also indications that direct funding—in contrast to tax incentives—could have a decreasing effect on public investment per patent in the long run.

M. Korber (✉) • M. Paier
Innovation Systems Department, AIT Austrian Institute of Technology GmbH, Donau-City-Straße 1, 1220 Vienna, Austria
e-mail: manuela.korber@ait.ac.at; manfred.paier@ait.ac.at

N. Gilbert et al. (eds.), *Simulating Knowledge Dynamics in Innovation Networks*, Understanding Complex Systems, DOI 10.1007/978-3-662-43508-3_5,

5.1 Introduction

Direct funding (government grants for research projects) is a traditional instrument for fostering innovation. It is considered adequate for fostering innovation activities, especially regarding high-risk projects and those of elevated societal relevance. The focus lies on projects with high uncertainty regarding probability of success, which—if the research activities in these projects prove successful—are likely to reap high and socially sustainable profits (Aiginger et al. 2009a, pp. 6–7).

In line with the Barcelona/Lisbon strategies of the EU and also developments in other OECD countries, public research policy in Austria in the last few years has put increasing weight on indirect funding, i.e., tax incentives for research. While tax incentives are associated with low entry barriers and low levels of administrative effort for beneficiaries, windfall gains are likely. On the other hand, direct government funding is conceded stronger incentives and a higher steering effect (Falk et al. 2009, p. 7). Thus, both instruments, direct funding as well as tax incentives have to be considered complementary and also their interaction deserves sufficient attention (Aiginger et al. 2009b, p. 89). Despite this consensus among innovation science and policy, the impact of indirect funding and its interaction with direct funding instruments has not yet been sufficiently investigated.

The life sciences have been in the focus of Austrian research policy during the last two decades. A considerable amount of public funds was spent to stimulate research activities in the field (e.g. 2007: EUR 104 million, Austrian Council for Research and Technology Development 2009). In 2004, more than 5 % of the Austrian public research budget was invested in life sciences, covering all parts of the innovation system with a combination of generic and life sciences-specific instruments and a focus on education, research and fiscal policy[1] (Reiss et al. 2005, pp. 74–75). Institutional funding by the government is to a large extent absorbed by universities, while the non-profit research sector is small in international comparison. Direct funding includes government programs supporting research collaboration and also a more institutionalized form of cooperation between science and industry (so-called competence centres).

A brief synopsis of the Vienna life sciences cluster may start in the mid-1980s, when the foundation of a joint venture of Boehringer Ingelheim and Genentech (IMP 2011) sparked off dynamic activities in life sciences. In 1999, an Austrian biotech program was introduced which led to the setup of the LISA VR cluster initiative[2] in 2001 which serves as a coordination and information platform for the life sciences innovation system in the Vienna region (Life Sciences in the Vienna Region 2007, p. 7). Since 2003, the focus of research policy on the regional level

---

[1] Reiss et al. (2005, pp. 74–75) used historical data (1994–2002) on policy activities and national performance in biotechnology for the validation of the historical analysis and benchmarked data regarding biotech policies in the year 2004.

[2] Life Science Austria Vienna Region (LISA VR) cluster (Life Sciences in the Vienna Region 2007).

lies on life sciences (WWTF 2011). By 2007, 170 life sciences companies were located in the Vienna region and engaged around 11,000 employees. In addition, 4,300 academic researchers were active in the field of life sciences (Life Sciences in the Vienna Region 2007, p. 7). Today, the Vienna region is leading in oncology, immunology and neurobiology, three quarters of the total amount of agents operate in red biotechnology (Jörg et al. 2006, p. 8) and a large part of the organizations is active in analytical methods and services, diagnostics and diagnostic technologies, microbiology or pharmaceuticals (Austrian Life Sciences Directory 2009).[3]

Generally, Vienna's reputation as a science think tank is fairly better than its effective ability to exploit and commercialize research results (Reiss et al. 2003; Senker et al. 2000, p. 605). The cluster is still considered in its development, the lack of venture capital and expertise regarding commercialization of research are named as the main reasons (Cooke et al. 2007, p. 250). There are not enough risk funds which fulfil the specific requirements to finance research in life sciences, e.g. the long developmental periods for pharmaceutics. In addition, the relatively low number of really successful companies (Tödtling and Trippl 2007, pp. 351–357) and the missing global pharmaceutical player with Austrian roots are constraints for its rosy progress.

The life sciences in Vienna may therefore be considered as an industry at the transition from a science-driven to an industry-driven research intensive sector. Still, direct funding exists at national as well as at regional levels, but it increasingly includes measures supporting research collaboration between science and industry. However, following the increasing relevance of indirect funding measures in Austria, the life sciences are deeply affected by this changing policy regime.

The lack of data and the necessity for ex-ante evaluation of policy measures call for new approaches to assessing the effects of direct funding, tax incentives and their potential interdependence. Agent-based modelling (ABM) techniques can be used to analyse systems with a large number of interacting agents and emergent system properties that cannot be deduced by aggregating the agents' properties (Axelrod and Tesfatsion 2006, pp. 74–75). So far, ABM has not yet been used for ex-ante impact analyses in Austrian RTI policy (Reiner and Smoliner 2012, p. 59). Nevertheless, the method can contribute quantitative scenarios to the political process of programme definition and design. One scenario with a specific intervention, e.g. the grant of public funds for research, can be compared to another scenario where *no* intervention is undertaken. Core prerequisites for this task are a clear model design and an empirical calibration to the situation under scrutiny. The simulation approach goes beyond simple extrapolations of historical data, as it is usually the case in the interpretation of evaluations. Moreover, the simulation avoids bias from strategic behaviour which might be relevant in ex-post evaluations. (Korber 2012, p. 73).

[3] Self-reported research fields of life sciences organizations in the Vienna Region; Source: Data adapted from Austrian Life Sciences Directory (2009).

In a previous paper (Korber et al. 2009) an agent-based model of the life sciences innovation system was conceptualized. Now we present an implementation of the model involving data on the Vienna life sciences sector that serves as a tool for policy makers to evaluate ex ante different public research funding alternatives. Inspired by the SKIN model (*Simulating Knowledge Dynamics in Innovation Networks*) (Ahrweiler et al. 2004; Gilbert et al. 2001; Pyka and Saviotti 2002), the agent-based model presented here includes research organizations and universities as additional agent types. Further agent heterogeneity is assured by introducing new agent attributes. For the first time, the kene concept (Gilbert 1997) is extended by the use of research fields and core competencies with empirically calibrated meaning. Thus, a more descriptive modelling strategy is chosen which involves close empirical embedding throughout the modelling process. We face a trade-off between closeness-to-reality and analytic clarity. Beyond scientific value, credibility of the model is guaranteed only if policy makers recognize the economic structure in detailed models and find them more reliable than abstract mathematical models characterized by consistency requirements of rationality and equilibrium (Fagiolo and Dawid 2008, pp. 351–352). Thus, the empirical calibration and validation of the agent-based model become a core issue.

This agent-based model of life sciences can be considered as an experimental laboratory for important questions in innovation and technology policy, because the anticipated effects of a single policy measure can be extracted before the particular funding action is taken. In this laboratory, alternative measures can be tested regarding their effectiveness and concerning their desired or undesired side-effects. It helps to reduce the arbitrariness of public interventions and to select a portfolio of instruments in order to achieve public goals through the most effective allocation of public research funds. In the model, the research performance of an agent and the system reveals the necessities for governance and government interventions. By sustaining the systemic view on innovation, agent-based models allow for short as well as long-term analyses of research results triggered by public research funds. The whole research process can be monitored starting from a project grant that leads to an invention and, further, tracking how this invention possibly is launched on the market. An agent's state and behaviour are monitored in every time-step and the time lag between a stimulus and the consequences in agent behaviour can be easily bridged by following the agent's actions after having received public funds. This facilitates also apprehending negative reaction behaviour (Korber 2012, pp. 72–73).

The chapter is organized as follows: Sect. 5.2 introduces the model at a glance, followed by the micro-foundation of the agents in Sect. 5.3. Then the macro-foundation of agent behaviour, interactions and regularities is given in Sect. 5.4. Section 5.5 provides information regarding how the system performance is measured. Section 5.6 focuses on the implementation of the policy experiments in the model, i.e. the role of public funding in the model, before Sect. 5.7 presents the simulation results. The chapter closes with a short discussion of the findings and an outlook on future research (Sect. 5.8).

5.2 Model Specification

This section provides a short description of the model (Korber 2012). Life sciences in Vienna is regarded as a sectoral innovation system (Malerba 2002) that consists of organizations involved in the creation, development and utilization of the sector's technologies and products. Processes of interaction and cooperation in technology development, competition and selection in innovative and market activities form the relations within the system (Breschi and Malerba 1997, p. 131). By including different types of agents and their strategies, this new model draws on previous research (Gilbert et al. 2001, 2007), in particular on the SKIN model (*Simulating knowledge dynamics in innovation networks*) developed by Gilbert et al. (2007) that focuses on market interaction and knowledge exchange among firms. The model of life sciences departs from previous research in several aspects. *First*, explicitly public sector research, such as universities and public research organisations, and different types of public funding is taken into account, and *second*, it is focused on analysing the effects of policy intervention in a localized life sciences innovation system.

While the SKIN model represents a reductionist approach (Axelrod 1997), this model attempts to integrate empirical detail (Hassan et al. 2008). This modelling strategy relates to models which emphasize the characteristics of specific industries. By relying on work by Malerba and Orsenigo (2002, p. 667), a case-based[4] model of the life sciences innovation system in Vienna is developed which is sufficiently detailed in terms of time and space. Knowledge-related processes and political interventions regarding knowledge production and exchange are at the centre of the model.

The agent-based model of life sciences is depicted in Fig. 5.1 with its specific input (financial resources) and output (research results) factors. The life sciences innovation system is viewed as a black box and the objective is to shed light on parts of the non-linear processes within this system. Although financial resources and research results are part of the system, they are extracted and illustrated separately as input and output boxes. Available data provides insights on these figures. There is no linear relationship between the input and the output of the system, but research on the—still largely unknown—on-going processes in the black box will make it easier to provide reliable policy advice. All these specific factors are outlined in the next sections. First, the agents are described with their specific attributes and strategies, and then the specific assumptions that govern their knowledge interactions are provided. Although these assumptions have been chosen based on careful considerations, the postulates in the model may differ from real practice where empirical references and theoretical evidence lack.

---

[4] A case-based model refers to an empirically circumscribed phenomenon, integrates rich empirical data and disposes of a high level of target and model details (Boero and Squazzoni 2005, pp. 8–10).

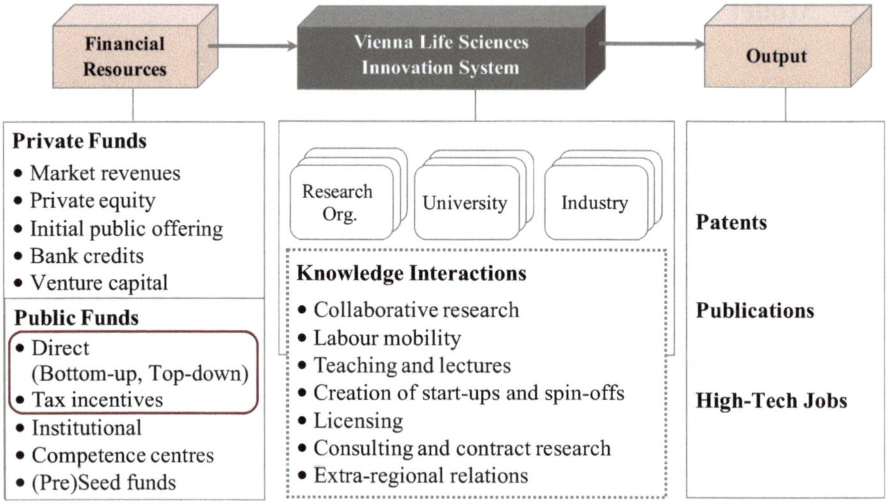

Fig. 5.1 The agent-based model

5.3 Micro-foundation of the Agents

For modelling the innovation system, we distinguish between exogenous and endogenous factors. Governmental authorities (Edquist 1997, p. 2) and financial organisations as important elements of an innovation system are modelled as exogenous entities that allow for public funds, credits or venture capital for the agents' investments. The agent-based model distinguishes three types of agents: university agents, research organisation agents and industry agents. Additional heterogeneity is modelled by different attribute values for each particular agent type as given in Table 5.1.

Behavioural dispositions are characterised by specific knowledge endowments and other attribute values that govern the exchange processes among agents. For later ease in expressing the knowledge interactions and the evaluation of research concepts by means of the fitness function, a selection of attributes is described by a specific notation. Each agent a_i ($i = 1, \ldots, I$) owns a set of kenes k_i (Gilbert 1997, pp. 8–10) representing its specific knowledge endowment K_i , given by

$$K_i = \{(r_m, c_n, \gamma_{imn}) | \gamma_{imn} > 0, m = 1, \ldots, M, n = 1, \ldots, N\} \text{ and } i \in \{1, \ldots, I\} \quad (5.1)$$

The kenes k_i of a_i are elements of the following set:

$$k_i \in \{(r_m, c_n, \gamma_{imn}) | \; m = 1, \ldots, M, n = 1, \ldots, N\} \text{ and } i \in \{1, \ldots, I\} \quad (5.2)$$

where m denotes a research field, n a core competency, and γ_{imn} an expertise level reflecting the experience and know-how gathered in the r_m and c_n (Pyka et al. 2002,

Table 5.1 Agent attributes

Attribute	Scale type	Notation	Value
ORGANISATION TYPE	Categorical		Industry, University, Research organisation
RESEARCH FIELD	Categorical	r_m	1, ..., 35
CORE COMPETENCY	Categorical	c_n	1, ..., 6
Expertise	Ordinal	γ_{imn}	1, ..., 10
FINANCIAL STOCK	Ratio		
EMPLOYEES	Ratio		
RESEARCHERS	Ratio		
FOUNDATION YEAR	Categorical		
RESEARCH ORIENTATION	Trichotomous	o_i	No research, Basic research, Applied research
Research attitude	Dichotomous		Incremental, Radical
Research strategy	Dichotomous		Go-it-alone, Collaborative
Partner search strategy	Dichotomous		Conservative, Progressive
Collaboration strategy	Dichotomous		Imitative, Collective

Note: Attributes in uppercase letters in the "Attribute" column (e.g. ORGANISATION TYPE) are empirically calibrated
Source: Austrian Life Sciences Directory (2009); Bureau van Dijk (2010), Aurelia database, complemented by own inquiry

p. 173). While the research fields *m* (e.g. dermatology, oncology, proteomics, etc.) of an agent refer to a scientific, technological field or a particular business domain (see Table 5.2), core competencies *n* (e.g. R&D, sales, service, etc.) relate to particular competencies within the specific research field as outlined in Table 5.3. In contrast to Pyka et al. (2002), research fields are defined in terms of categorical rather than metric variables.

Apart from their research fields in scientific, technological and business domains, agents are characterised by core competencies, as displayed in Table 5.3. Both, research fields and core competencies are measured in terms of nominal variables. Every agent reaches a certain expertise level γ_{imn} within each r_m of its research fields combined with a particular core competency *n*. This expertise level γ_{imn} indicates the acquired know-how in the particular technological capacity over the time-steps in the course of the simulation (Pyka et al. 2002, p. 173).

Finally, agents are not only characterised by these knowledge endowments, but also by other attributes, as entitled in Table 5.1, that are widely viewed to be crucial for agent behaviour. The activities of an agent are calibrated by its empirically observed attributes and dispositions. Whether an agent's *research orientation*[5] is basic or applied determines the agent's behaviour during individual and collaborative research. The *research strategy* of an agent decides, if an agent performs only individual research or if it heads for cooperation.

[5] This is an improvement of the attribute *research direction* in the SKIN model which is not empirically based and randomly determined (Gilbert et al. 2007, pp. 102–103).

Table 5.2 Specification of the agents' research fields

r_m	Research field in a scientific, technological or business domain	r_m	Research field in a scientific, technological or business domain
1	Analytical methods & services	18	Lab equipment, medical & surgical equipment
2	Antibodies	19	Metabolomics
3	Bacterial & viral diseases/Antiinfectives	20	Medical technology & devices
4	Cardiovascular diseases	21	Microbiology
5	Cell & tissue culture	22	Nanobiotechnology
6	Clinical research & tests	23	Neurobiology/Neurodegenerative diseases
7	Consulting	24	Nutrition/Food/Feed
8	Dermatology	25	Oncology
9	Diagnostics/Diagnostic technologies	26	Pharmaceuticals
10	Drug development/Drug delivery	27	Plant breeding & genetics
11	Environmental issues	28	Proteomics
12	Enzymology/Protein engineering/ Fermentation	29	Process technology
13	Gene & cell therapy, viral vectors	30	Regenerative medicine
14	Genomics	31	Services (synthesis, sequencing, spectroscopy)
15	Immunology/Allergology	32	Stemcells
16	Industrial processing	33	Structural biology
17	Informatics in the life sciences	34	Vaccines
		35	Veterinary activities

Note: r_m denotes research field m; the measurement scale ranges from 1 to 35
Source: Austrian Life Sciences Directory (2009)

Table 5.3 Specification of the agents' core competencies

c_n	Core competency within a particular research field m
1	R&D
2	Contract research
3	Production & processing
4	Sales
5	Service
6	Education & training

Note: c_n denotes core competency; the measurement scale ranges from 1 to 6
Source: Austrian Life Sciences Directory (2009)

In the model, agents that pursue *collaborative research* may follow two different strategies. Collaboration might be realized according to an imitative or a collective strategy. While the *imitative* strategy excludes individual research and agents perform only collaborative research, the *collective* collaborative strategy comprises in-house as well as joint research. Agents following the imitative strategy search constantly for collaboration partners and focus only on imitation (Ahrweiler et al. 2004; Gilbert et al. 2001; Pyka et al. 2002, p. 176). Further determinants of an agent's possibilities are *financial stock, number of employees* and *research staff*.

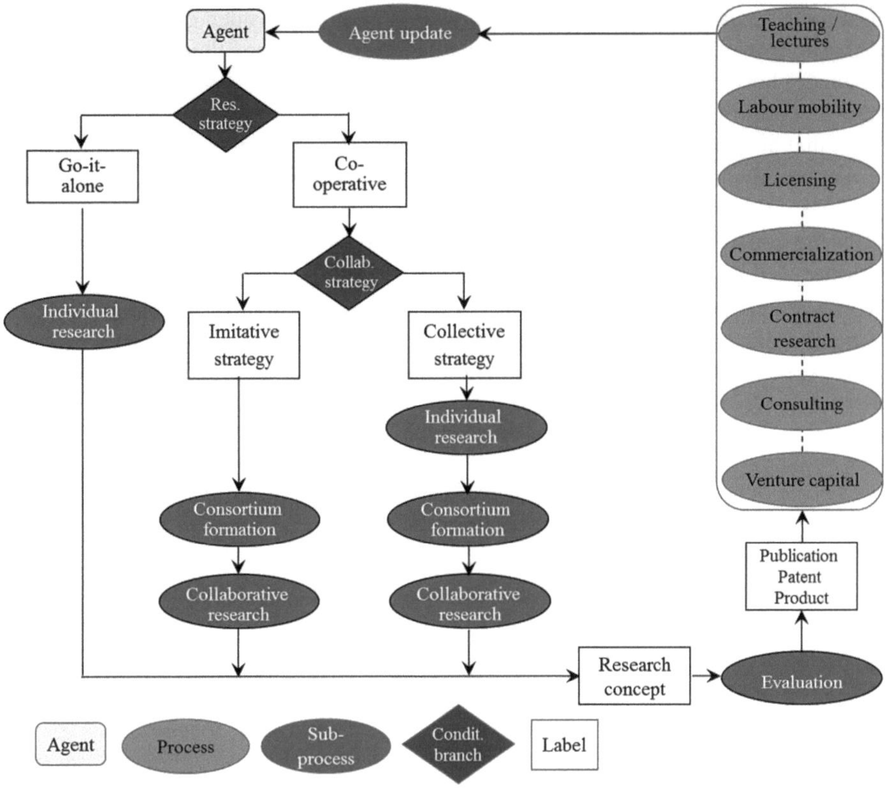

Fig. 5.2 Main agent procedures during one time-step

5.4 From Micro to Macro: Behaviour, Interactions and Regularities

At the beginning of each time-step, agents decide in accordance to their strategy whether they perform exclusively individual research because they follow the go-it-alone strategy or they go for cooperation and form consortia (Pyka et al. 2002, p. 176). Figure 5.2 illustrates the main agent procedures during one time-step (which corresponds to a quarter of the year). During the simulation, an agent is assumed to be able to modify or expand its kene set K_i through individual research efforts or cooperation with other agents in joint research activities. No matter whether carried out alone or in collaboration, research is costly on the one hand, but, on the other hand it leads to the acquisition of new research fields and core competencies for the agent's kenes.

During learning and *research* processes, agents acquire new research fields according to their *research attitude*, whereby *incremental* research is associated with the acquisition of rather similar research fields, and *radical* research with more dissimilar research fields.[6] Previous research (e.g. Ahrweiler et al. 2004; Pyka et al. 2002) treats research fields formally as integer numbers, thus the dissimilarity (distance) between two research fields $r_1 = k$ and $r_2 = l$ can be described simply by their algebraic difference $|k - l|$. In contrast, our model employs agents' research fields that are empirically based, and are therefore represented not by a numeric but by a nominal variable. Accordingly, a metric in the research field space (a similarity measure on the set of $n = 35$ research fields) is defined as follows: Two research fields are considered similar, if they are both held by many agents (co-occurrence). A concept of proximity is reached by taking the number of co-occurrences $Y_{r_1 r_2}$ of two respective research fields r_1 and r_2 in the agent population and calculating the respective Jaccard-Index (Leydesdorff 2008) as a measure of thematic proximity of these research fields. Let r_1 and r_2 denote research fields, that is $r_1, r_2 \in R = \{r_m | m = 1, \ldots, M\}$, then the Jaccard index S is given by

$$S_{r_1 r_2} = \frac{Y_{r_1 r_2}}{Y_{r_1} + Y_{r_2} - Y_{r_1 r_2}} \tag{5.3}$$

where

$$Y_{r_1} = \sum_{\substack{r_1 = 1 \\ r_2 \neq r_1}}^{M} Y_{r_1 r_2}, \quad Y_{r_2} = \sum_{\substack{r_2 = 1 \\ r_1 \neq r_2}}^{M} Y_{r_1 r_2} \tag{5.4}$$

and $Y_{r_1 r_2}$ is the number of co-occurrences of research fields in organizations. The index ranges from [−1, +1]; the higher the index the higher the similarity between two research fields. By this, we approximate a knowledge space which allows determining similarity between research fields (Korber and Paier 2011, p. 607). Agents that operate according to the *incremental* research attitude choose the new research field with the highest similarity to the nearest research field of the agent, while those with a *radical* research attitude take the research field with the lowest similarity to the nearest research field of the agent. On the one hand, agents with a radical research attitude investigate entirely diverse market opportunities and choose a new research field (r_m) which is thematically very far from the research fields they hold currently. On the other hand, incrementally researching agents select a research field which is very close and face lower risks in integrating the new research field. The new core competency is selected according to the majority of the core competencies that the particular agent holds. The expertise level of the respective kene is set to the beginner-level, which is one. The new kene k_i which consists of the new research field m, core competency n and expertise level γ_{imn} is

[6] The concept of incremental and radical research behaviour was already introduced in the SKIN model, but implemented without empirical calibration and based on random choice (Gilbert et al. 2007, pp. 102–103).

attached to the agent's knowledge endowment. The financial stock is diminished by the costs for research (see Table 5.6 in the Appendix). If the kenes of the agent's knowledge endowment are used for current research and a research concept is formed during the particular period, the expertise level γ_{imn} is increased by one. In case that research is performed by a partnership or a network, the research attitude held by the majority of the participating agents is chosen. Incremental and radical research performed by partnerships and networks are subject to the same process.

Research fields are *forgotten* as fast as they are acquired. Inactivity in research is punished and research fields which are not used for current research (to create the research concept during a certain quarter) suffer declining expertise levels. The kene's expertise level γ_{imn} is decreased by one, until eventually it may drop to zero. Consequently, this kene k_i is eliminated from the agent's knowledge endowment K_i (Pyka et al. 2002, p. 174). Once all kenes are forgotten, the agent exits the system. The same is valid for learning by interacting, i.e., only knowledge which is actively used by the agents in a partnership or a network, and a research concept is created, increases an agent's knowledge base.

It is further postulated that the processes of learning by doing, learning by using and learning by interacting (Andersen and Lundvall 1997, p. 254) lead to the generation of *inventions*.[7] Agents use their knowledge characteristics to contribute to the creation of inventions which in the model are called *research concepts*.[8] After individual research (see Fig. 5.2), each agent a_i selects three elements from its knowledge endowment K_i that denotes the research concept D_i. These are seen as key competencies and are chosen without replacement with a probability proportional to the expertise level γ. Thus D_i is a set of distinct kenes $k_1^{'}$, $k_2^{''}$, and $k_3^{'''}$

$$\begin{aligned} D_i &= \left\{k_i^{'}, k_i^{''}, k_i^{'''}\right\} \\ k_i^{'}, k_i^{''}, k_i^{'''} &\in K_i \\ k_i^{'} &= \left(r_m^{'}, c_n^{'}, \gamma_{imn}^{'}\right) \\ k_i^{''} &= \left(r_m^{''}, c_n^{''}, \gamma_{imn}^{''}\right) \\ k_i^{'''} &= \left(r_m^{'''}, c_n^{'''}, \gamma_{imn}^{'''}\right) \end{aligned} \tag{5.5}$$

As in every innovation system, particularly in life sciences, the interdependencies among agents are manifold. The interaction of agents reflects the mechanisms of cooperation and competition. Institutions shape their actions and are either formal, such as intellectual property rights, or informal, e.g. established practices (Edquist and Johnson 1997, p. 46). Research cooperation in Austrian life sciences takes place in various ways: The most intensive knowledge flows between

[7] An invention is a new idea before its commercialisation (Fischer 2003, p. 344).

[8] The research concept is inspired by the innovation hypothesis used by Pyka et al. (2002, pp. 174–178).

these agents are associated with the conduct of joint research that may lead to co-authored patents and academic papers.

Self-organised consortia form the basis of *collaborative research* which can be enhanced by public funding if specific policy requirements are fulfilled. As is supported by the model, collaborative research fosters an easier access to new knowledge than *individual research* efforts of the agents. Agents that follow the *cooperative research strategy* seek to engage in consortia with adequate partners. According to the *conservative partner search strategy* similar potential partners are preferred while according to the *progressive partner search strategy* dissimilar organisations are preferred (Gilbert et al. 2007, p. 103). In this compatibility check, the similarity of two potential partners is measured by the number of research fields they have in common (Gilbert et al. 2007, p. 103; Pyka et al. 2002, pp. 177–178). Additionally, cooperation experience is taken into account since past collaboration partners are preferred over other partners (Pyka and Scholz 2008, pp. 6–13). If both agents agree, they enter a consortium. A consortium $G_{ibi)}$ to which a_i belongs is defined as follows:

Let $G_{ibi)} \subseteq A$ where $|G_{ibi)}| = H_{ibi)}$ where $H_{ibi)}$ is the number of agents in $G_{ibi)}$ which are chosen in the partner search procedure according to the compatibility check. Additionally $bi) \in \{1,\ldots,B_i\}$, where B_i is the total number of consortia where a_i is a member. $K_{ibi)}$ is the union of the research concepts D_h of the agents a_h in the consortium $G_{ibi)}$ where a_i is a member.

$$K_{ib(i)} = \bigcup_{a_h \in G_{ib(i)}} D_h \tag{5.6}$$

The research concept $D_{ibi)}$ of a consortium $G_{ibi)}$ is defined as three distinct kenes $k'_{ibi)}, k''_{ibi)}$ and $k'''_{ibi)}$ of the knowledge endowment $K_{ibi)}$ of the consortium $G_{ibi)}$. The learning by interacting procedure is similar to learning by doing. Success or failure feeds back on the agents' kene sets. Kene sets are merged in the following way: Each research field in the other agent's research concept that is new to the agent is integrated in its kene set and the core competency that dominates its current activities is attributed. Due to the fact that integration of new, external knowledge is difficult, the expertise level of the respective research field is downgraded to one (Pyka et al. 2002, p. 178). For each kene that is not new, if the other's expertise level is higher, its expertise level is adopted, otherwise the expertise level remains unchanged (Gilbert et al. 2001, p. 6).

In addition to collaborative research, agents are linked due to various other knowledge interactions that are delineated in the following. Knowledge spillovers[9] occur through labour mobility, adjunct faculty or as a result of the creation of spin-offs, etc.

Figure 5.3 outlines the relations that are considered between the main types of agents. The model puts special emphasis on knowledge production and exchange

[9] Knowledge spills over from the generating agent to other agents and there is a smaller compensation than the value of the knowledge, or even none (Fischer 2003, p. 345).

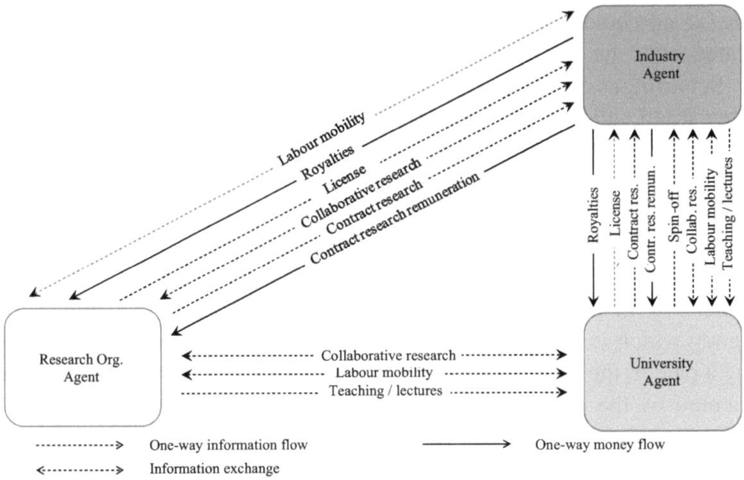

Fig. 5.3 Agent types and their relations

processes among the agents. These processes are realized by the modification and exchange of kenes, which is in some cases also compensated by money flows.

Labour mobility occurs mainly between the science and the industry sector, or through the hiring of university graduates by companies or research organisations. In Vienna's life sciences sector, less labour mobility is observed within the industry sector, at least at the local level and major importance is attributed to international labour mobility (Tödtling and Trippl 2007, p. 361). Labour mobility is reflected in the model, as researchers change their affiliation and one randomly picked kene is eliminated from one agent's knowledge endowment and integrated into another randomly chosen agent's knowledge endowment. If this kene is new to the other agent, it is simply added to its kene set. For each research field (and its associated core-competency) that is not new, the higher expertise level is chosen and attributed to the respective kene and the new kene is integrated into the new affiliation's kene set. This transfer on the labour market feeds back on the organisations' number of researchers and employees—the original organisation's figures are diminished by one and the respective figures are increased by one for the new organisation.

Adjunct teaching is very common in the sector, since life sciences specialists and managers often give lectures in educational organisations. This channel is a rather formalized interaction type triggering personal contact and possibly the transfer of tacit knowledge. In the model, through *teaching and lectures*, knowledge of life sciences experts is passed on to other organisations. In these instances one randomly chosen kene is copied and transferred to another agent. If the respective research field is not yet known by the other agent, it is added to its knowledge endowment together with its core competency and its expertise level. If the research field and its associated core competency is not new to the agent only the expertise level is updated. In case the lecturer's expertise level is greater than that of the other agent which listened to the lecture, the expertise level of the lecturer is adopted.

Licensing agreements link companies to other agents that own patents, and are thus related with the transfer of explicit knowledge which requires less personal contact (Schartinger et al. 2002, p. 305). Research organisation agents, university agents and small and medium-sized enterprises that filed a patent may license it to large enterprises that are possible licensees. The number of potential licensees is limited to a randomly chosen number below 10. Potential licensees decide in favour of the patent with the best quality (defined by the sum of expertise levels in the research result). The research fields of the licenser's research concept are integrated in the licensee's knowledge endowment. Following the idea that in-house core competencies, e.g. R&D,[10] are easily adapted to other research fields, the core competency which is owned most is chosen for the new kenes. After successfully licensing a patent, the expertise level of the licenser is raised and the financial stock is augmented by the royalties. Royalties may also be subject to price adjustment mechanisms.

Further knowledge flows arise through consulting and contract research that agents with basic or applied research orientation perform for other agents. In the model, agents may perform *contract research* or *consulting* for every other agent. The only exception are agents with the research orientation $o_i =$ no research. They can offer only consulting services and not contract research. Similar to the market mechanism described below, the quality of the research concept is given by the sum of its expertise levels. Agents decide based on the quality and price of the research result. Contract research and consulting clients integrate the research fields of the consultant's or contract researcher's result in their kene set. The performance of contract research or consulting for other agents may not only provoke knowledge flows but affects the financial stock as well. Consequently, money flows between the agents occur. In exchange, the financial stock of the contract researcher or consultant is increased by the remuneration. If the price adjustment mechanism is activated, the remuneration depends on the demand for this agent's research output in the current year.

Apart from the above mentioned interactions, also other knowledge interactions take place, but are not included in the model. These comprise synergies because companies use research infrastructure and university facilities, or rely on academic expertise and buy prototypes which have been developed at universities. Less formalized types of knowledge interactions come from sabbatical periods, the joint supervision of master and PhD theses, the employment of graduates by industry agents, and the training of firm members. An intense transfer of tacit knowledge, without any formal agreements, occurs during conferences, informal meetings, or due to the reading of publications and patents which create common knowledge in a certain field (Schartinger et al. 2002, p. 305).

Extra-regional relations in the model are represented in a very stylized manner. 87 per cent of the agents are cooperating extra-regionally (Heller-Schuh and Paier 2009, p. 162: EU projects), if their market status in the on-going year was seller,

---

[10] R&D is short for Research and Development.

consultant, contract researcher or licenser. A randomly generated kene is included into the agent's kene set and the financial stock is increased after extra-regional interactions.

5.5 Measuring the System Performance

Agents (or, in case of collaborative research, groups of agents) create *research output* from their *research concepts*. It is distinguished between three types of research output: *scientific papers, patents* (Pyka and Scholz 2008, p. 10) or *products*. Whether a research output is a scientific paper, a patent, or a product is determined by the research orientation of the producer. The fitness of a research concept D_i after individual research is defined by the fitness function f as follows:

$$f_i = \begin{cases} 1 & \text{if} \left((I-1)\widetilde{f}_i / \sum_{\substack{u=1 \\ u \neq i}}^{I} \widetilde{f}_u \right) \geq 1 \\ 0 & \text{otherwise} \end{cases} \tag{5.7}$$

where

$\widetilde{f}_i = \gamma'_{imn} + \gamma''_{imn} + \gamma'''_{imn}$ and $\gamma'_{imn}, \gamma''_{imn}$ and γ'''_{imn} are the expertise levels of the corresponding kenes of a_i in D_i, i.e., the sum of the expertise levels $\gamma'_{imn}, \gamma''_{imn}$ and γ'''_{imn} in the research concept D_i,$\widetilde{f}_i$, is compared to the average sum of the expertise levels in the research concepts of all other research concepts $\widetilde{f}_u$. If the sum of the expertise levels in the research concept D_i of agent a_i is equal or above compared to the average of the rest of the agent population, this research concept is considered successful.

Depending on the result of this evaluation phase a certain status is attributed to each research concept. The transformation of research concepts into research outputs works as follows: If $f_i = 1$, then D_i is considered successful and is transformed into outputs according to the following decision rule which is based on the research orientation o_i of the particular agent a_i:

$$o_i = \begin{cases} 1 = \text{no research} \\ 2 = \text{basic research} \\ 3 = \text{applied research} \end{cases} \tag{5.8}$$

We define: If the agent's research orientation o_i is "no research", then the output of the research concept is measured in terms of "commercialized on the market". If o_i = "basic research", the output of the research concept is measured in terms of a "scientific publication". If, o_i = "applied research", the output of the research concept is measured in terms of a "patent application". The successful generation of output feeds positively back on the agents: By increasing the expertise levels γ_{imn}

of the kenes of a successful research concept and incorporating these updated kenes into the kene sets of the involved agents, the agents cumulatively increase their expertise levels.

The knowledge endowment of a consortium is the union of the knowledge endowments of all agents that participate in the cooperative project. The evaluation of the research concept of project groups is determined by the same fitness function *f* as given by Eq. (5.7). The majority of research orientations in the consortium defines the output that is generated after the cooperative research project, see Eq. (5.8). If the research concept is considered successful, the type of output is determined by the research orientation o_i which is held by most of the agents in the consortium. If there is no majority of agents that hold the same research orientation o_i, the output is attributed randomly.

Performance measurements also affect new firm generation. A particularly successful and profit-making incumbent attracts start-ups. This reflects the generation of variety as well as the diffusion of economically relevant know-how. So as to represent the lack of experience and initial capital, the start-up's expertise level and financial stock are low (Gilbert et al. 2007, pp. 103–104). While at the beginning a start-up company is dependent on public research funding, later it is able to attract private investors as well. The creation of spin-off companies represents a particular knowledge flow linking academia with the business world. University members hold company stakes or create start-up companies (Fischer et al. 2006, p. 138). Sometimes spin-off companies also grow out of companies or research organisations, consequently, facilitating knowledge flows. In the model, after a successful period (i.e. the agent's success status is set to "success"), agents may create a *start-up* or *spin-off* organisation. The new agent is chosen randomly from the empirical input file according to its foundation year and added to the population. The first 12 years during the simulation, agents with a foundation year until 2010 are chosen from the empirical data set. After the simulation period equivalent to 2010 the pool for the calibration of new agents is the whole data set based on the 128 agents and their foundation year corresponds to the current foundation year in the simulation. The kene set of the new agent consists of the successful agent's research concept, but the expertise levels of the new kene set are restricted to three. The financial stock of the start-up is set to nearly the same amount as the founder's financial stock is diminished (see Table 5.6 in the Appendix). If the research attitude of the founder and the entrant is radical, the financial stock is increased by public (pre)seed funding (see Table 5.8 in the Appendix).

Therefore, we can say that even the phenomenon of system-level learning is implemented in the model. Organisations that continuously perform research activities improve their expertise levels and acquire new research fields. Consequently, they are more successful in creating research outputs and attract start-up and spin-off organisations which focus on these particular research fields. Therefore, positive dynamics are sparked off and the life sciences innovation system benefits as a whole (Gilbert 2008, p. 16).

Agents *exit* the system because they either forget all their kenes or they go bankrupt. If every kene of an agent's knowledge endowment has been forgotten (i.e. the expertise level of every kene of an agent has dropped to zero and was deleted), the agent is excluded from the agent population due to its inactivity. Another reason for agent exit is bankruptcy. If the financial stock of an agent results zero or less, then this agent exits the agent population. University agents and public research organisation agents are excluded from this rule because they cannot go bankrupt. If they have a financial stock below zero, their loss is compensated by government. It is assumed that the negative value (loss) is multiplied by a factor of -1.5, thus achieving a positive financial stock again.

Several possibilities to acquire funds are the consequence of successful research. If they were granted a patent or if they did not perform individual research (research orientation $o_i =$ no research), but were evaluated successfully, agents may commercialise their innovations on the *market*. The quality of the stylized innovation is defined by the sum of the expertise levels in the research concept. A maximum of ten other agents are chosen randomly that are potential buyers. The market then evaluates the research concepts. If the innovation offers the cheapest price with the best quality in comparison to the competitors, the innovation is bought by other agents. The financial stock of the seller is increased and the buyer's is decreased. In exchange, the buyer receives the research fields. If a research field in the seller's research concept is new to the buyer, it is added to the buyer's kene set and the core competency is selected the one that the buyer is specialized in (i.e. that it holds predominantly), the respective expertise level is set to one. For each research field (and its associated core competency) that is not new, if the seller's expertise level is greater than the buyer's, its expertise level is adopted. Of course the price for innovations depends on the market conditions. If the price adjustment mechanism is activated, the price depends on the demand for this seller's research output in the current year. If the seller has more than four clients in the current period, the price rises by 10 % and if not, the price decreases by 10 %.

If they attract investors, industry agents and private research organisations might receive *private equity*, and SMEs can go public (see Table 5.6 in the Appendix). In this case they make an *initial public offering* (*IPO*), i.e. it is assumed that their financial stock is increased by the total share price[11] and their organisation type is changed into an LE. Companies and private research organisations with incremental research attitude and conservative partner search strategy that are in need of money, i.e. their financial stock is below EUR 5,000 can take out a loan from the *bank*. SMEs with a radical research attitude might receive *venture capital* with a percentage chance of 23 % (Jörg et al. 2006, p. 35).

[11] This is done due to simplification reasons, although being aware of the fact that an IPO does not in every case lead to an increase in share capital, i.e. an increase in the financial stock respectively. It might also be the case that a previously private stake is offered at the stock exchange without an increase in share capital.

Additionally to the number of patent applications and academic papers, the creation of high-tech jobs is monitored (see the output side of the black box in Fig. 5.1). Due to the fact that patenting reflects the organisation's expectations which are often not fulfilled, it is not enough to rely on the number of patents as a unique indicator. As every indicator is generally arguable and restricted, a complementary indicator is introduced, the creation of high-tech jobs. Whereas the number of patents and academic papers are to a high degree specific and arbitrary, this system variable allows measuring the performance from a dynamic perspective at a macro-level and reflects the potential for future output. It results from agents' attributes, such as the number of research personnel employed by the organisations as well as the expertise levels specifying their research fields, or the agents' turnover. The creation of high-tech jobs reflects not only the sector's performance over time but it also represents the benefits of public research funding efforts for the national economy. Therefore, accumulated values of several attributes (e.g. the creation of start-ups) can be surveyed. Given that academic papers and patents reflect conducted research, these indicators together with the creation of high-tech jobs characterise the development of the life sciences innovation system.

5.6 The Role of Public Research Funding

The model allows simulating different public research funding regimes and, facing the complexity of the life sciences innovation system, emphasis is laid on the role of public research funding in the system. Public funds comprise institutional funding granted specifically to science agents, whereas programme and project funding goes to science as well as to industry agents. In a recent analysis of research networks in the Vienna life sciences sector, 136 projects in eight funding programmes were identified. Out of this number, two programmes are European, namely the *Life Quality* programme in the 5th EU framework programme as well as *Medical and Biotechnology* in EUREKA. The national funding activities comprise the Austrian NANO initiative, the GEN-AU Genome Research Austria (GEN-AU 2009) in addition to five specific competence centres. To be emphasized here is the fact that Viennese organisations are largely involved in European projects (87 %), and less at a national (6 %) or regional (7 %) level (Heller-Schuh and Paier 2009, p. 162).

In the case of Austrian life sciences, various types of joint research are considered as fruitful although one of the major drawbacks is red tape, i.e., the involvement in bureaucratic and non-research activities. Generally, it is often criticized that the funding system in Austria is too complex and confusing, and that for some research stages (e.g. clinical research phase 2) funding is not provided at all. Specifically, the non-existence of standardized contracts, fundraising, and accounting for funding institutes claim considerable time which could be used for core business (life-science.at 2008). Public research policy at the regional level

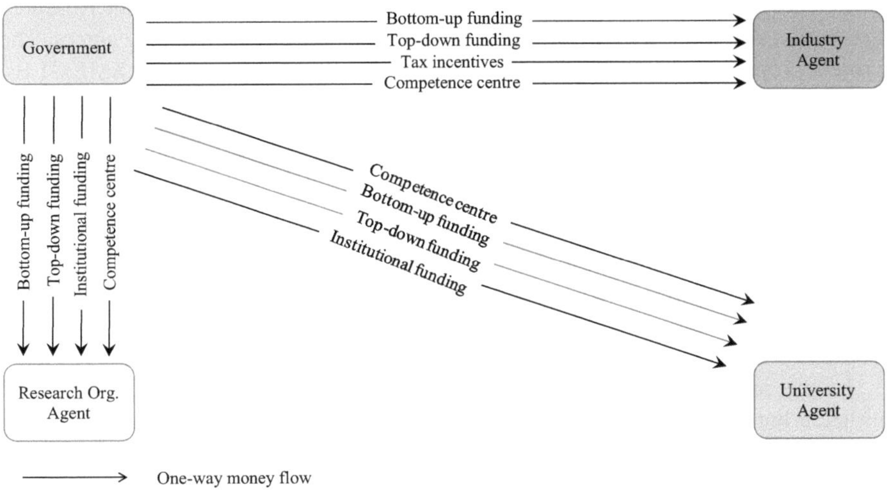

Fig. 5.4 Public research funding as a financial resource

aims to improve possibilities offered to resident companies and organisations regarding access and use of funding support, and promote regional innovation potential (Cooke 2002, p. 133). For a localized sectoral innovation system like the Vienna life sciences, it is important how effective public interventions are in the creation of sustainable dynamics within the cluster and its relations with the outside world.

This agent-based model allows the analysis and comparison of the effects of different funding types in a localized life sciences innovation system regarding research performance. Hereby, the various types of direct funding—with or without requirement for inter-organisational cooperation—can be compared to indirect funding (tax incentives) and also to the case of inexistent policy intervention.

Government funding for research with the requirement to cooperate triggers the structure of collaboration networks which influences the agent-specific knowledge output in a dynamic way. This belief has governed research policy throughout Europe in the last decades, and it continues to do so also at the regional level. As illustrated above in Fig. 5.4, public research funding realizes money flows from the *government* component to industry, university and research organisation agents.

Top-down direct funding includes grand challenges or specific topics that government seeks to promote more intensively. According to empirical data, industry agents that hold certain research fields and core competencies are more likely to receive funds within the ZIT calls (The Technology Agency of the City of Vienna – ZIT 2010). *Bottom-up direct funding* is available for individual research of industry agents as well as for collaborative research projects by all agents. A certain share of the research costs is refunded to the agents, thus, the financial stock of the respective agents is increased (see Table 5.8).

Science-industry relations in the model are funded by the Austrian Research Promotion Agency (FFG) through the implementation of *competence centres*. The project proposals are screened and analysed with respect to how many SMEs, LEs, research organisation agents and universities are in the consortium. Based on the FFG criteria it is decided whether a consortium can apply as one of the three types of competence centres K, K1 and K2. Depending on its members (FFG defines them to contain a certain number of SMEs, LEs and university as well as research organisation agents) and the type of competence centre, the share of public funds to cover project expenses and the project duration are determined (The Austrian Research Promotion Agency – FFG 2010). After the end of the respective project duration, the competence centres are dissolved and project partners are memorized as previous project partners.

In the model, government provides not only funding for programmes and projects but also university agents and public research organisation agents with *institutional funding*. The financial stock of public research organisations and university agents is increased by an empirically calibrated funding volume each quarter. Moreover, if universities or public research organisations face bankruptcy and their financial stock is in the red, it is assumed that their minus is compensated by government.

Indirect funding comprises tax allowances or the deduction of research expenses from tax. Since 2011, a research premium of 10 % is granted on each industry agent's research costs until an annual ceiling amount of EUR 100,000 per agent is reached. The research premium can be claimed only after the end of each economic year, for simplification in the model, this is done at the end of the last quarter of each year (Legal Information System of the Republic of Austria (RIS) 2012).

A further important measure to stimulate cluster development is the promotion of start-ups and entrepreneurship (Tödtling et al. 2006, p. 14). Start-ups and young SMEs are drivers of innovation dynamics in life sciences (Audretsch 1995; Feldman 2000; Fuchs and Krauss 2003, p. 5; Giesecke 2000). Therefore, *preseed and seed financing* are public funding tools implemented in the model. If the research attitude of a start-up and its founder is radical, both receive (pre)seed funding.

In contrast to indirect research funding which benefits all industry agents as long as they perform research, direct funding exerts a stronger governance effect by taking the knowledge endowments of submitted proposals into account. It is thus expected to steer the innovation process with regard to the direction of research at the agent as well as at the system level. Government knows the once submitted or published kenes of all agents and serves somehow as an autonomous agent which makes funding decisions. This concept covers also the process of lobbying by agents. Thus, governmental intervention improves to some extent the *innovation oracle* that has been used in other ABMs of innovation (Pyka et al. 2002, pp. 175–181).

Table 5.4 The scenarios at a glance

Scenarios	Policy measure	Public funds: Parameter settings referring to Direct funding (bottom-up, top-down)	Public funds: Parameter settings referring to Tax incentives
(i) Direct funds	Only *direct* public research *funds* are granted	X	
(ii) Tax incentives	Only *tax incentives* for research are provided		X
(iii) Direct funds & tax incentives	Both funding measures are simultaneously available	X	X

Note: X indicates that the specific intervention is switched *on* during the simulation runs

5.7 Simulation Results

This section introduces three distinct funding scenarios (Table 5.4) and illustrates corresponding simulation results. These scenarios refer (i) to direct funds only, (ii) to tax incentives only, and (iii) to the reference scenario with both direct funds and tax incentives at work. Detailed parameter settings are provided in Tables 5.5, 5.6, 5.7, and 5.8 in the Appendix. Due to spatial constraints, the output parameters are presented on the system level only (i.e. the total number of patent applications, scientific publications and high-tech jobs).

The three scenarios differ only regarding the availability of *direct funds* and *tax incentives*. They are considered *ceteris paribus*, i.e. all other parameters in Fig. 5.1 referring to *Private Funds* (Market revenues, Private equity, Initial public offering, Bank credits, Venture capital) and *Knowledge Interactions* (Collaborative research, Labour mobility, Teaching and lectures, Creation of start-ups and spin-offs, Licensing, Consulting and contract research, Extra-regional relations) as well as the *Public Funds* institutional funding, competence centres and (pre)seed funds are switched on.

The time step t_0 in the simulation corresponds to the beginning of the year 1999 and to the empirical agent structure and number according to the empirical database at that time, i.e. at the set-up the simulation features 74 agents. The modelling experiment was carried out for 120 time steps, corresponding to a period of 30 years. The plots all refer to the respective values at the year end. Ten different simulation runs were carried out for each experiment while sweeping the random seed values (i.e. the pseudo randomness in the model). The simulation results presented here show the mean values of these ten simulation runs for each scenario.

The development of the yearly amount of public funds provided in each of the three scenarios is given in Fig. 5.5. Public funds in the simulation model start with empirically calibrated funding requirements and levels. The direct funding Scenario (i) involves considerably higher amounts of public funds, while the Scenario (ii) which bases on tax incentives only requires less public funds. Figure 5.5 further underlines clearly that we face a complex system, as the two Scenarios (i) and

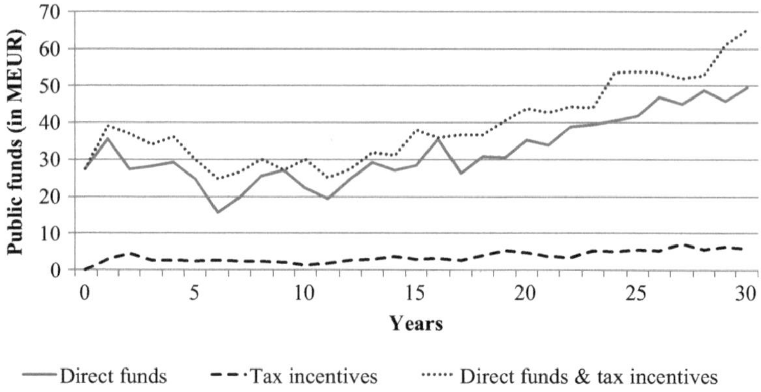

Fig. 5.5 Public funds per specific intervention(s)

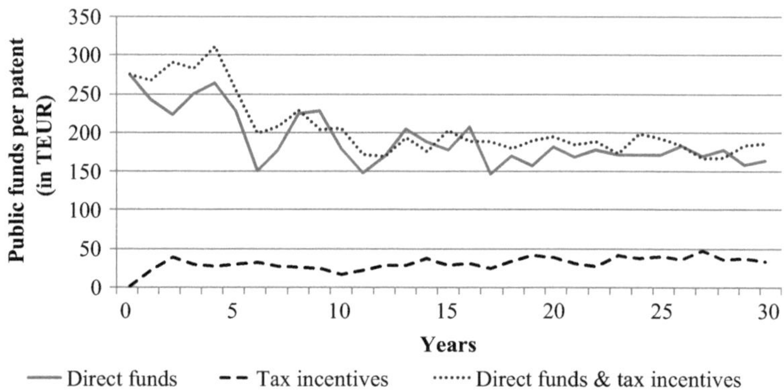

Fig. 5.6 Public funds per patent

(ii) do not add up to the evolution observed in Scenario (iii). Scenario (iii) emerges at a slightly higher level than the scenario with direct funds.

In our results we focus on efficiency considerations from a public policy point-of-view, i.e. we relate output variables to public funds injected into the system. Figures 5.6, 5.7 and 5.8 show the evolution of public expenditures per patent, publication and high-tech job in the three scenarios. As shown in Fig. 5.6, the amount of direct public funds per patent is considerably higher than the tax incentives that need to be provided in order to generate one successful patent. Thus, tax incentives seem to be more efficient in stimulating patents. The most expensive scenario in terms of public funds per patent is the Scenario (iii) where direct funds for research projects as well as tax incentives are provided. Nevertheless, one has to keep in mind that the beneficiaries of tax incentives are firms only. This explains the different level of direct funds which flow into the system.

Fig. 5.7 Public funds per publication

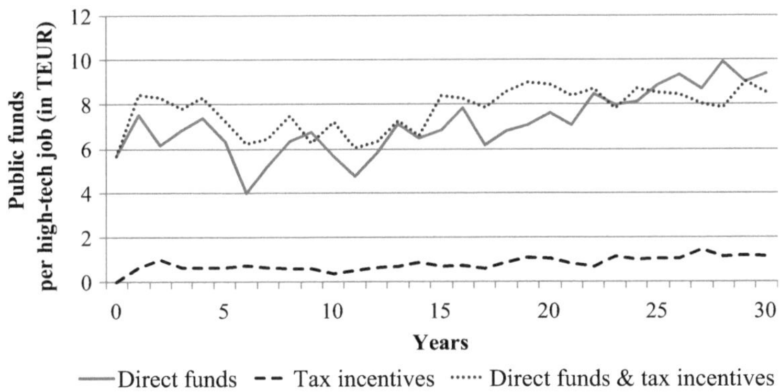

Fig. 5.8 Public funds per high-tech job

Interestingly, we observe a slight increase in public spending per patent in the case of tax incentives only, Scenario (ii), which means that patents become more expensive in the long run if not complemented by direct funds. In contrast, public spending per patent shows a decreasing long-run tendency in Scenario (i). A possible interpretation for this phenomenon might be the positive effect of direct funds for knowledge transfer from universities and research organisations to firms. A lack of collaboration with universities requires a financial compensation by firm's individual research which is more expensive than collaborative research.

In contrast to patenting, we observe a different tendency for publication in the Scenarios (i) and (iii). Figure 5.7 shows that public spending per publication slightly increases in the long run in both scenarios with direct funds.

The public funds per high-tech job (Fig. 5.8) show upward slopes for all scenarios as well, especially in Scenario (i) with direct funds only.

Summing up, the tax incentives only scenario, Scenario (ii), seems to be the most efficient one in terms of public funds per output indicator. This is due to the fact that tax incentives are not available for the whole agent population but only for firms. The scenarios with direct funds involved are more volatile and exhibit a considerably higher level of required public funds, since all agent types are eligible for public funding in this case.

5.8 Conclusions and Outlook

This chapter develops an agent-based model of a life sciences innovation system that draws upon considerable empirical detail. Existing agent-based modelling approaches focus on network formation and innovation processes in the life sciences sector in general by taking into account the complexity of a specific innovation system. Our model advances from former models in several aspects. (1) The elaborate calibration of the model is expected to improve the validity of the model, especially in a technology and innovation policy context. With this model the dynamic effects of different modes of public funding in several respects can be explored. (2) A unique database on the life sciences Vienna innovation system was constructed that has been used for empirical calibration. (3) The model consists of agents with empirically calibrated knowledge endowment (kenes) that are changed and recombined through individual and joint research and different agent interactions. Thus, the dynamic distribution of research fields and core competencies is modelled, allowing for the creation of novelty in the system. (4) This model uses a similarity index that imprints a metric on the research field space. (5) The model is also of particular value for conducting policy experiments due to the inclusion of system parameters such as the creation of high-tech jobs, scientific publications and patent applications. By this, system-wide innovative output depending on an evaluation process with corresponding fitness functions can be monitored. (6) This model also allows studying the role of cooperation within the innovation system and across its borders and investigating how collaboration intensity is related to innovative output. All these features render the model very useful for policy advice.

We focus on an intensely debated issue of Austrian research policy, i.e. the effects of tax incentives in comparison to direct public funds. In order to focus on this question, three scenarios are contrasted, i.e. Scenario (i), where only direct funds but no tax incentives are accessible, Scenario (ii) with tax incentives but without the availability of direct public research funds, and a reference scenario where all agents benefit from the whole research funding portfolio and thus, from both funding measures simultaneously (Scenario iii). The simulation results are presented with a special focus on the output indicators patents, publications and high-tech jobs on the system level.

The study suggests that direct funding requires a considerably higher volume of public funds compared to tax incentives. In order to get an idea about the efficiency of both funding measures, the paper presents the amount of public funds that are required for a patent application, a scientific publication or the creation of a high-tech job. Tax incentives are more efficient in promoting the total number of patents, scientific publications and high-tech jobs. This might indicate that the main value-added from direct public funds is to be expected in the governance of research specialisation or its direction.

Inherent limitations of this simulation model are given by its overall complexity. Consequently, further testing and model validation should be carried out with statistical analysis. Next steps in our research include an analysis of the stated research objective only for the firms in the agent population which would reveal more significant statements on efficiency measures. Further potential for improvement of the model includes the regional dimension by focusing on the impact of extra-regional cooperation on regional knowledge output. Moreover, the evolution of the variety or specialization of the agent population will be investigated. The use and further development of this simulation model in an innovation policy context will be guided by involving policymakers, thus relying on the companion-modelling approach, also referred to as expert validation.

Acknowledgements This chapter reports results of research carried out in the framework of the Innovation Economics Vienna – Knowledge and Talent Development Program. The first author gratefully acknowledges the scholarship provided by this program. Furthermore, the first author would like to thank her doctoral supervisors Manfred M. Fischer, Andreas Pyka and Michael Barber for their valuable contributions in excellent discussions. Part of the research was also funded by the Austrian Science Fund through the project "Innovation networks for regional development" (FWF-DACH I886).

Appendix

The presented model is programmed with the software tool NetLogo, version 5.0 (Wilensky 1999). The specific program code for the NetLogo model on which this paper is based is available from the author on request. The simulation runs described in Sect. 5.7 are based on the parameter settings given in Tables 5.5, 5.6, 5.7 and 5.8.

Table 5.5 Agent parameter settings (all three scenarios)

Parameter	Value
Initial agent population	75 agents: agent number and structure according to life sciences organisations in the Vienna region as per 1999 (Austrian Life Sciences Directory 2011)
Organization type; Differentiated org. type	Industry (SME, LE), University (university, university of applied sciences), Research organization (public, private): according to life sciences organisations in the Vienna region as per 1999 (Austrian Life Sciences Directory 2011), Aurelia (Bureau van Dijk 2010), organisation (business reports, web pages, etc.)
Research fields	Calibrated according to life sciences organisations in the Vienna region as per 1999 (Austrian Life Sciences Directory 2011), complemented by own inquiry
Core competencies	Calibrated according to life sciences organisations in the Vienna region as per 1999 (Austrian Life Sciences Directory 2011), complemented by own inquiry
Expertise level	Uniform random distribution from 0 to 9
Financial stock	According to yearly turnover or budget of particular organisations (Austrian Federal Ministry for Science and Research 2008, p. 62; Bureau van Dijk 2010), organisation (business reports, web pages, etc.)
Employees	Austrian Life Sciences Directory (2011), Aurelia (Bureau van Dijk 2010), organisation (business reports, web pages, etc.), complemented by own inquiry
Researchers	Austrian Life Sciences Directory (2011), Aurelia (Bureau van Dijk 2010), organisation (business reports, web pages, etc.), complemented by own inquiry
Foundation year	Austrian Life Sciences Directory (2011), Aurelia (Bureau van Dijk 2010), organisation (business reports, web pages, etc.), complemented by own inquiry
Research orientation	No research, Basic research, Applied research (Austrian Life Sciences Directory 2011), Aurelia (Bureau van Dijk 2010), organisation (business reports, web pages, etc.), complemented by own inquiry
Share of agents	67 % incremental research attitude 34 % go-it-alone research strategy (perform own research) 49 % conservative partner search strategy 58 % imitative collaboration strategy (during cooperation)

Table 5.6 System parameter settings and empirical calibration (all three scenarios, per quarter)

Parameter	Value	
Random seeds	20, 25, 30, 40, 45, 65, 70, 75, 80, 90	
Receipts and expenditures		
Costs of own research	Research organization agents	EUR 100,207 multiplied by total no. of researchers of the agent
	University agents	EUR 32,702 multiplied by total no. of researchers of the agent
	Research expenses scientific sector and number of employees 2007 (Schibany et al. 2010, pp. 143 and 149)	
	Industry agents	
	Total no. of employees <50, per quarter	EUR 25,543 multiplied by total no. of researchers of the agent
	Total no. of employees 50–249, per quarter	EUR 44,220 multiplied by total no. of researchers of the agent
	Total no. of employees $\geq$ 250, per quarter	EUR 43,250 multiplied by total no. of researchers of the agent
	Research expenses business sector and number of employees 2007 (Schibany et al. 2010, p. 140)	
Costs of cooperative research per partner	Uniform random distribution from	EUR 0–15,000
Labour mobility	Probability of 3 %	
Royalties	Uniform random distribution from	EUR 0–600,000
Remuneration for consulting projects	Uniform random distribution from	EUR 0–60,000
Remuneration for contract research	Uniform random distribution from	EUR 0–60,000
Remuneration for extra-regional relations	Uniform random distribution from	EUR 0–10,000
Private equity	Uniform random distribution from	EUR 0–25,000
	Probability per quarter of (Schibany et al. 2010, p. 141)	5 %
Bank credit	Uniform random distribution from	EUR 0–100,000
	SME: Uniform random distribution from	EUR 0–70,000

(continued)

Table 5.6 (continued)

Parameter	Value	
Initial financial stock of start-ups	Minimum nominal capital for Austrian GmbH: EUR 35,000 (§ 6 GmbHG)	
	LE: Uniform random distribution from	EUR 0–100,000
	Minimum share capital for Austrian AG: EUR 50,000 (§ 7 AktG)	
	University: Uniform random distribution from	EUR 0–20,000
	Research org.: Uniform random distribution from	EUR 0–20,000
Revenue per sold innovation	Uniform random distribution from	EUR 0–100,000
Venture capital	Venture capital is limited to investment in seed, start-up, early development and expansion stages. Later stage replacement and buy-out investments are excluded. Venture capital investment data for the life sciences for 2007, Average size per investment per year: USD 791,600: as of 20 January 2012: EUR 611,524 (OECD 2009, pp. 96 and 101)	
Cash flow from issued shares	EUR 100,000	
SMEs go public	Probability of 0.1875 % (Schibany et al. 2010, p. 141)	

Table 5.7 Switches to determine the agents' option for action (all three scenarios)

Parameter	Parameter settings
Private funds	
Market revenues	ON
Private equity	ON
Initial public offering	ON
Bank credits	ON
Venture capital	ON
Knowledge interactions	
Collaborative research	ON
Labour mobility	ON
Teaching and lectures	ON
Creation of start-ups and spin-offs	ON
Licensing	ON
Consulting and contract research	ON
Extra-regional relations	ON
Agent exits	
Exit due to bankruptcy	ON
Exit because expertise is forgotten	ON

Table 5.8 Empirical calibration of government funds in the three scenarios

Parameter	Parameter settings	Scenario (i)	Scenario (ii)	Scenario (iii)
Direct funding	Bottom-up direct funding: Share of funded project expenses: Uniform random distribution from 0 to 50 % Top-down direct funding: ZIT Calls: calibrated volume and probability referring to particular research fields and core competencies (The Technology Agency of the City of Vienna – ZIT 2010)	ON	OFF	ON
Indirect funding	Depends on research expenses, 10 % research premium to industry agents, ceiling amount: 100,000 per agent per year (Legal Information System of the Republic of Austria (RIS) 2012)	OFF	ON	ON
Institutional funding	Institutional funding volume per quarter: Uniform random distribution from 0 to 86,953 (Univ. report 08: Austrian Federal Ministry for Science and Research 2008, p. 62)	ON	ON	ON
Science-industry relations: creation of competence centres	Calibrated volume, probability and duration (FFG criteria: K (p. 38), K1 (p. 22), K2 (p. 12): The Austrian Research Promotion Agency – FFG, 2010)	ON	ON	ON
(Pre)Seed funding	Calibrated volume of (pre)seed funds; Average volume per funded project = 484,800; Uniform random distribution from 0 to 947,000 (AWS funding portfolio: Austria Wirtschaftsservice 2011)	ON	ON	ON

References

Ahrweiler P, Pyka A, Gilbert N (2004) Simulating knowledge dynamics in innovation networks. In: Leombruni R, Richiardi M (eds) Industry and labor dynamics: the agent-based computational economics approach. World Scientific, Singapore, pp 284–296

Aiginger K, Falk R, Reinstaller A (2009a) Reaching out to the future needs radical change – towards a new policy for innovation, science and technology in Austria; evaluation of government funding in RTDI from a systems perspective in Austria; summary. WIFO, Vienna

Aiginger K, Falk R, Reinstaller A (2009b) Reaching out to the future needs radical change – towards a new policy for innovation, science and technology in Austria; evaluation of government funding in RTDI from a systems perspective in Austria; synthesis report. WIFO, Vienna

AktG. Aktiengesetz 1965. Current version, § 7, Gesetz 9

Andersen E, Lundvall B-Å (1997) National innovation systems and the dynamics of the division of labor. In: Edquist C (ed) Systems of innovation: technologies, institutions and organizations. Pinter/Cassell Academic, London, pp 242–265
Audretsch DB (1995) Innovation and industry evolution. MIT Press, Cambridge, MA
Austria Wirtschaftsservice (2011) Seedfinancing portfolio – life sciences
Austrian Council for Research and Technology Development (2009) Strategie 2020, 19 August 2009. http://www.forschungsstrategie.at/. Retrieved 3 June 2010
Austrian Federal Ministry for Science and Research (2008) University report 2008. 2nd corrected edition
Austrian Life Sciences Directory (2009) http://www.lifesciencesdirectory.at/. Retrieved 13 June 2009
Austrian Life Sciences Directory (2011) http://www.lifesciencesdirectory.at/. Retrieved 3 Dec 2011
Axelrod R (1997) The complexity of cooperation. Agent-based models of competition and collaboration. Princeton University Press, Princeton, NJ
Axelrod R, Tesfatsion L (2006) Appendix A: a guide for newcomers to agent-based modeling in the social sciences. In: Tesfatsion L, Judd KL (eds) Handbook of computational economics: agent-based computational economics, vol 2, 13th edn. Elsevier, Amsterdam, pp 1647–1659
Boero R, Squazzoni F (2005) Does empirical embeddedness matter? Methodological issues on agent based models for analytical social science. J Artif Soc Soc Simulat 8(4):6
Breschi S, Malerba F (1997) Sectoral innovation systems: technological regimes, Schumpeterian dynamics, and spatial boundaries. In: Edquist C (ed) Systems of innovation: technologies, institutions and organizations. Pinter, Cassell Academic, London, pp 130–156
Bureau van Dijk (2010) Aurelia company database. http://www.bvdep.com/tw/AURELIA.html. Retrieved 8 Aug 2010
Cooke P (2002) Regional innovation systems: general findings and some new evidence from biotechnology clusters. J Technol Transf 27(1):133–145
Cooke P, De Laurentis C, Tödtling F, Trippl M (2007) Regional knowledge economies: markets, clusters and innovation. Edward Elgar, Cheltenham
Edquist C (1997) Systems of innovation approaches – their emergence and characteristics. In: Edquist C (ed) Systems of innovation: technologies, institutions and organizations. Pinter/Cassell Academic, London, pp 1–35
Edquist C, Johnson B (1997) Institutions and organisations in systems of innovation. In: Edquist C (ed) Systems of innovation: technologies, institutions and organizations. Pinter/Cassell Academic, London, pp 41–63
Fagiolo G, Dawid H (2008) Agent-based models for economic policy design: introduction to the special issue. J Econ Behav Organ 67(2):351–354
Falk R, Neppl-Oswald E, Trebicka K, Weixlbaumer U (2009) Kohärenz des Instrumentenmix – Zusammenspiel der direkten und indirekten Forschungsförderung Systemevaluierung der österreichischen Forschungsförderung und -finanzierung. WIFO, Wien
Feldman MP (2000) Where science comes to life: university bioscience, commercial spin-offs, and regional economic development. J Comp Pol Anal 2(3):345–361
Fischer M (2003) The new economy and networking. In: Jones D (ed) New economy handbook. Academic, Elsevier, San Diego, CA, pp 343–367
Fischer MM, Schartinger D, Rammer C, Fröhlich J (2006) Knowledge interactions between universities and industry in Austria: sectoral patterns and determinants. In: Fischer MM (ed) Innovation, networks, and knowledge spillovers: selected essays. Springer, Berlin, pp 135–166
Fuchs G, Krauss G (2003) Biotechnology in comparative perspective. In: Fuchs G (ed) Biotechnology in comparative perspective. Routledge, London, pp 1–13
GEN-AU (2009) Genomforschung in Österreich. http://www.gen-au.at/index.jsp?lang=de. Retrieved 18 Mar 2009
Giesecke S (2000) The contrasting roles of government in the development of biotechnology industry in the US and Germany. Res Pol 29(2):205–223
Gilbert N (1997) A simulation of the structure of academic science. Sociol Res Online 2(2)

Gilbert N (2008) Agent-based models, vol 153. SAGE, Los Angeles, CA
Gilbert N, Pyka A, Ahrweiler P (2001) Innovation networks – a simulation approach. J Artif Soc Soc Simulat 4(3):8
Gilbert N, Ahrweiler P, Pyka A (2007) Learning in innovation networks: some simulation experiments. Phys A 378(1):100–109
GmbHG (2012) Gesetz über Gesellschaften mit beschränkter Haftung. § 6 Abs. 1 Satz 2 GmbHG
Hassan S, Antunes L, Pavon J, Gilbert N (2008) Stepping on earth: a roadmap for data-driven agent-based modelling. The Centre for Research in Social Simulation, Paper 5
Heller-Schuh B, Paier M (2009) Regional – national – europäisch: Wiener F&E-Netzwerke aus der Mehr-Ebenen-Perspektive. In: Leitner K-H, Weber M, Fröhlich J (eds) Innovationsforschung und Technologiepolitik in Österreich: Neue Perspektiven und Gestaltungsmöglichkeiten. Studienverlag, Innsbruck, pp 154–179
IMP (2011) Research Institute of Molecular Pathology: IMP. http://www.imp.ac.at/the-imp/history/. Retrieved 25 Jan 2011
Jörg L, Endemann M, Streicher J, Rammer A, Hinze S, Roloff N, et al (2006) Life Science – Standort Wien im Vergleich, January 2006. http://www.lisa-vr.at/contentFiles/LS%20Studie%202006.pdf. Retrieved 5 May 2008
Korber M (2012) Agent-based modelling of complexity in life sciences – with a special emphasis on the impact of public funding on research activities. Book series of the Innovation Economics Vienna – Knowledge and Talent Development Programme PhD- & Master-Theses, Doctoral work, vol 16, AIT Austrian Institute of Technology GmbH, Vienna University of Economics and Business, PhD- & Master-Theses Series 16
Korber M, Paier M (2011) Exploring the effects of public research funding on biotech innovation: an agent-based simulation approach. Paper presented at the proceedings of the eighth international conference on complex systems "unifying themes in complex systems", Boston, Quincy, MA, USA, 26 June to 1 July 2011
Korber M, Paier M, Fischer MM (2009) An agent-based view of the biotech innovation system. Reg Direct Int Sci J 2(2):33–55
Legal Information System of the Republic of Austria (RIS) (2012) Einkommensteuergesetz EStG 1988, Prämien für Forschung und Bildung (Forschungsprämie, Bildungsprämie). § 108c
Leydesdorff L (2008) On the normalization and visualization of author co-citation data: Salton's cosine versus the Jaccard index. J Am Soc Inf Sci Technol 59(1):77–85
Life Sciences in the Vienna Region, 2009 (2007)
life-science.at (2008) Wie kooperieren etablierte Unternehmen und Universitäten erfolgreich? http://life-science.at/success/site/content/view/377/207/. Retrieved 21 July 2008
Malerba F (2002) Sectoral systems of innovation and production. Res Pol 31(2):247–264
Malerba F, Orsenigo L (2002) Innovation and market structure in the dynamics of the pharmaceutical industry and biotechnology: towards a history-friendly model. Ind Corp Change 11(4):667–703
OECD (2009) OECD biotechnology statistics – 2009. OECD, Paris
Pyka A, Saviotti P (2002) Innovation networks in the biotechnology-based sectors. In: Pyka A, Küppers G (eds) Innovation networks: theory and practice. Edward Elgar, Cheltenham, pp 75–107
Pyka A, Scholz R (2008) A narrative description of the agent based NEMO model. NEMO Working Paper, 11
Pyka A, Gilbert N, Ahrweiler P (2002) Simulating innovation networks. In: Pyka A, Küppers G (eds) Innovation networks: theory and practice. Edward Elgar, Cheltenham, pp 169–196
Reiner C, Smoliner S (2012) Outputorientierte Evaluierung öffentlich geförderter FTI-Programme – Möglichkeiten und Grenzen, Studie im Auftrag des Bundesministeriums für Verkehr, Innovation und Technologie March 2012, Zentrum für Soziale Innovation, ZSI, Joanneum Research, JR

Reiss T, Hinze S, Lacasa ID, Mangematin V, Enzing C, van der Glessen A et al (2003) Efficiency of innovation policies in high technology sectors in Europe (EPOHITE). Contract No. HPV1-CT-2001-00005, Final report

Reiss T, Lacasa ID, Mangematin V, Corolleur F, Enzing C, van der Giessen A, et al (2005) Benchmarking of public biotechnology policy. Fraunhofer Institute Systems and Innovation Research, European Commission Enterprise Directorate General, Final report

Schartinger D, Rammer C, Fischer MM, Fröhlich J (2002) Knowledge interactions between universities and industry in Austria: sectoral patterns and determinants. Res Pol 31(3):303–328

Schibany A, Berger M, Buchinger E, Dachs B, Dinges M, Ecker B, et al (2010) Life sciences in Österreich. Austrian Research and Technology Report 2010, pp 138–165

Senker J, Enzing C, Joly PB, Reiss T (2000) European exploitation of biotechnology-do government policies help? A recent survey of public spending on biotechnology in Europe suggests that money alone cannot stimulate growth of the sector. Nat Biotechnol 18(6):605–608

The Austrian Research Promotion Agency – FFG (2010) FFG Fokus: Kompetenz und Exzellenz – Die Comet-Zentren und -Projekte: Österreichs Flaggschiffe in der Spitzenforschung. Österreichische Forschungsförderungsgesellschaft mbH, 1st edn

The Technology Agency of the City of Vienna – ZIT (2010) Call archive. http://www.zit.co.at/en/funding/current-calls/call-archive.html. Retrieved 10 Mar 2010

Tödtling F, Trippl M (2007) Knowledge links in high-technology industries: markets, networks or milieu? The case of the Vienna biotechnology cluster. Int J Entrepren Innovat Manag 7(2/3/4/5):345–365

Tödtling F, Trippl M, von Gabain J (2006) Clusterentwicklung und -politik im Biotechnologiesektor Wien im Kontext internationaler Erfahrungen. SRE-Discussion Paper 2006/02, Research report, Vienna University of Economics and Business – Institute of Regional Development and Environment

Wilensky U (1999) NetLogo. http://ccl.northwestern.edu/netlogo. Retrieved 2 Mar 2012

WWTF (2011) Programme & Ausschreibungen – life sciences. http://www.wwtf.at/programmes/ls/. Retrieved 15 June 2011

Chapter 6
R&D Policy Support and Industry Concentration: A SKIN Model Analysis of the European Defence Industry

Fulvio Castellacci, Arne Fevolden, and Martin Blom

Abstract The European defence industry is a highly concentrated market. The European Union has recently introduced a new Directive that is supposed to ensure free trade in defence material and make the European defence industries more efficient and competitive. This chapter makes use of a SKIN model to analyze how this on-going process of EU liberalization is likely to affect industry concentration in the defence industries of large and small European countries. The simulation analysis points out two main results: (1) There is a clear connection between market liberalization and a reduction in industry concentration. (2) The effect of the liberalization reform on industry concentration is clear for both large and small EU economies, but it is stronger for larger countries. On the whole, these results emphasize the importance of introducing market liberalization in such a highly concentrated and protected market as the defence industry.

F. Castellacci (✉)
Norwegian Institute of International Affairs, C.J. Hambros plass 2D Pb 8159 Dep, 0033 Oslo, Norway
e-mail: Fulvio.Castellacci@nupi.no

A. Fevolden
NIFU Nordic Institute for Studies in Innovation, Research and Education, P.O. Box 5183, Majorstuen, 0302 Oslo, Norway
e-mail: arne.fevolden@nifu.no

M. Blom
Center for Technology, Innovation and Culture, University of Oslo, P.O Box 1108, Blindern, 0317 Oslo, Norway
e-mail: martin.blom@gmail.com

N. Gilbert et al. (eds.), *Simulating Knowledge Dynamics in Innovation Networks*, Understanding Complex Systems, DOI 10.1007/978-3-662-43508-3_6,

6.1 Introduction

The European defence industry is a highly concentrated market. This is to a large extent the result of an extensive and wide-ranging use of protectionist policies—such as discriminatory procurement practices and offset (counter-trade) requirements. Regardless of country size, specialization pattern and GDP per capita, the defence sector in European countries is almost always dominated by a few oligopolistic defence contractors. But, as several commentators have pointed out, this industry structure comes at a price (Edwards 2011). The high level of industry concentration and the low integration among the domestic markets of EU countries have forced economies in this area to accept more costly and less efficient defence equipment, which hampers the international competitiveness of this industry *vis-à-vis* other regions in the world.

This weak competitiveness has recently prompted the European Union (EU) to introduce a new Directive (*The European Union's Defence and Security Procurement Directive, 2009/81/EC*) that is supposed ensure free trade in defence material and make the European defence industries more efficient and competitive. Although the actual impact of this Directive is still uncertain, it has already spawned a wide-ranging debate about how the different Member Countries will be affected, and whether it will actually succeed in creating a more open and competitive European defence sector (Edwards 2011; Blom et al. 2014). This chapter wants to contribute to this debate by analysing how this on-going process of EU liberalization is likely to affect industry concentration in the defence industries of large and small European countries.

Industry concentration measures the level of competition within an industry and thereby the possibility of incumbent companies to extract monopoly rents. A reduction in industry concentration may have positive effects such as lower prices for consumers and a higher general level of welfare for society. In recent research within Schumpeterian economics, industry concentration has also been associated with the technological characteristics of industries, such as technological opportunities, cumulativeness and appropriability conditions, and the properties of the knowledge base (Klepper 1996; Malerba and Orsenigo 1996). It has been common in this literature to distinguish between two types of 'technological regimes': *Schumpeter Mark I* which describe highly fragmented industries with an open, unstable, competitive structure where incumbents often engage in radical and discontinuous types of innovations; and *Schumpeter Mark II* which describe highly concentrated industries with a stable oligopolistic structure where incumbents mostly engage in incremental and cumulative types of innovation.

The defence industry, due to its oligopolistic and highly concentrated structure, most closely resembles the description of a Schumpeter Mark II type regime. It is reasonable to argue that the introduction of the EU Directive might reduce these industries' concentration, and hence have wide-ranging impacts. The EU Directive might not only increase competition and lower prices, it might also increase the mortality rate among European defence contractors and alter the defence industries' rate and direction of innovation. Motivated by these issues, this chapter will try to answer two questions:

(i) To what extent does a more open market and reduced policy protection affect industry concentration in the European defence industries?
(ii) Are there important differences in the ways the defence industries of larger and smaller European countries will be affected by the liberalization process?

To investigate these questions, this chapter will set up and analyse an agent-based model of the defence industry, which will simulate the effect of the EU Directive on the dynamics of the European defence industries. The model is based on and extends the SKIN model (*Simulating Knowledge Dynamics in Innovation Networks*), which provides an analysis of private firms' interactions and knowledge dynamics in high-tech industries and were developed in a number of recent articles by Gilbert et al. (2007), Pyka et al. (2007) and Ahrweiler et al. (2011). Our approach is rooted in these recent models, but it extends them further by applying them to a study of industry concentration within a context of European defence sector liberalization. Specifically, the model we present here is based on, and extends further, the recent work of Blom et al. (2014), focusing on how this type of SKIN model can shed new light on the relationship between R&D policy changes, on the one hand, and industry concentration, on the other.

The simulation analysis carried out in this chapter looks at the effect of the EU Directive by analysing changes along four "liberalization" parameters: increases in firms' collaboration propensity, greater degree of competition they face in the domestic market, and changes in the criteria for allocation of R&D policy support, with a consequent reduction in the amount of policy support that defence based companies receive. These four parameters represent in a stylized manner the effects of more open markets and reductions in policy support that might follow from the implementation of the EU Directive. The simulation analysis points out two main results: (1) There is a clear connection between market liberalization and a reduction in industry concentration. (2) The effect of the liberalization parameters on industry concentration are clear for both large and small countries, but the effect on larger countries is more pronounced. On the whole, these results emphasize the importance of introducing market liberalization in such a highly concentrated and protected market as the defence industry.

The chapter is organized as follows. Section 6.2 points out some key stylized facts about the industry concentration in the European defence industry; Sect. 6.3 presents the model; Sect. 6.4 defines the firm-level and industry-level variables used in the simulation analysis; Sect. 6.5 analyzes six alternative policy scenarios; and Sect. 6.6 concludes by summarizing the key results and implications of the work.

6.2 Industry Concentration in the European Defence Industry

Although the ambition of the new EU Directive is to create an open and highly integrated defence market, the situation today is that the European defence sector consists of a collection of well-protected and largely independent national defence industries. This is to a large extent the result of the extensive and wide-ranging use of defence industrial policies that most European countries have—at least up until recently—engaged in (Brauer and Dunne 2004; Markowski et al. 2010). Most European countries have made use of some form of offset requirements, where they have demand that foreign defence contractors should make purchases from their national defence industries in exchange for contracts to deliver defence equipment to them. Most European countries have also engaged in some form of discriminatory procurement practices, where they have awarded their national defence companies contracts regardless of whether foreign suppliers could have provided cheaper or better equipment. And most European countries have supported their national defence companies through the use of R&D contracts, which they have awarded to national companies based on the firms' importance for economic development or national security.

The European defence industries are also highly concentrated on downstream (front-end) activities. The literature on the defence sector points out that the production of defence equipment happens through a long value chain—where companies take part in different stages of the production process. Closest to the customer, we find the *primes* that are in charge of complete weapon platforms (e.g. aircrafts, submarines and frigates), at the next stage, the *systems integrators* that are in charge of weapon systems (e.g. missiles and command and control systems) and at the last stage, the *component supplier* that are in charge of components and sub-systems (e.g. cables and explosive charges). Among the European countries, the primes are mainly located in the largest countries—such as UK (BAE Systems), France (Thales), Germany (Thyssen Krupp) and Italy (Finmeccanica)—and they usually do not number more than one or a few in each country. The systems integrators are more numerous in the larger countries, and there are a few of them in smaller countries such as Norway (Kongsberg) and Finland (Patria). The component suppliers are usually small and medium-sized companies, and they are fairly numerous in most European countries. Since the primes and systems integrators have a very dominant position (large market shares) in the European defence sector, the overall industry concentration in most European countries is considered to be very high.

Although precise information on industry concentration in the defence sector is not available, the few studies that have been conducted have found that industry concentration is high and growing. A study conducted by Kuschel (2008)—which used the Herfindahl-Hirschman Index (HHI) to calculate the concentration of the defence industries in the UK, Germany, France, Italy and Sweden—found that all of these countries had an industry concentration that qualified as 'highly concentrated,' in the vernacular of the U.S. Federal Trade Commission and the U.S. Department of Justice. The same study also concluded that the industry concentration among the largest European defence producing nations were significantly higher than in the US, a country which has been considered to have a very concentrated industry (Kuschel 2008). Other studies conducted by the Stockholm International Peace Research Institute (SIPRI) have found that the largest European defence contractors' share of the market has been large and has increased in the last two decades, making the European defence industries even more concentrated (http://www.sipri.org).

While the industrial concentration is high on average in all European countries, it might be even higher in the smaller economies of the region. In one of the few direct comparisons that have been carried out, Kuschel (2008) found that the industry concentration in Sweden was higher than in any of the larger European countries, apart from Italy. Anecdotal evidence suggests that the largest companies in smaller European countries such as Norway, Finland and Switzerland have a disproportionally large share of the market, with companies such as Kongsberg, Patria and RUAG ranking as the 59th, 87th, 70th largest defence contractors in the world (http://www.sipri.org, statistics for 2011). Considering that the defence industries in these smaller European countries are significantly smaller than in the larger European countries, this would indicate that they have an even higher industry concentration.

Since the EU Directive is now in a process of being transposed into national laws, it is still an open question to what extent it will succeed in opening up the European defence industries and reduce their industry concentration. The intention behind the Directive is to ensure free trade of defence and security equipment within the European Economic Area, and it is likely that many Member States will find it more difficult to maintain defence industrial policies such as discriminatory procurement practices (national favouritism), offset (counter-trade) requirements and R&D contracts. But even if the EU Directive is completely successful and European defence contractors can compete uninhibited for contracts across Europe, the European defence industries will arguably still be fairly concentrated. Calculations made by Kuschel (2008) found that even a fully open European defence market would be considered "moderately concentrated" in the vernacular of the U.S. Federal Trade Commission and the U.S. Department of Justice. Thus, the achievement of an open and perfectly competitive European defence industry will depend not only on the removal of existing protectionist policies, but also on a different dynamics in the European defence industry, where the smaller defence contractors should play a larger and more prominent role than they have had until now.

6.3 The Model

Our analysis of industry concentration in the European defence industry is based on the SKIN model (*Simulating Knowledge Dynamics in Innovation Networks*), introduced in a number of recent papers by Gilbert et al. (2007), Pyka et al. (2007) and Ahrweiler et al. (2011). This is an agent-based model analyzing firms' interactions and knowledge dynamics in knowledge intensive industries. This chapter is rooted in this recent model and extends it further by applying it to the study of the defence sector. While the main description of the model follows the basic pillars of the SKIN approach, our model departs from it in some important ways, in the

Table 6.1 Our contribution to the SKIN model class

New aspect	Key feature	Specific characteristics
Focus and objectives	We focus on *export performance and liberalization*, incorporating new features from the firm heterogeneity and international trade literature into the SKIN model	*New parameters*: (1) Export threshold; (2) export intensity *New outcome variable*: (1) Export concentration ratio
Model mechanism	*Publicly-funded incremental innovation*: Successful firms may receive a public grant to finance a new R&D project	*New parameters*: (1) Success threshold; (2) product quality threshold; (3) competence breadth threshold; (4) public subsidy
Presentation	A more extended presentation of some of the characteristics of the model	*Formal presentation* of the model (Sect. 6.3) *Diagram* with overview of model's functioning (Fig. 6.1)

The model presented in this paper is largely based on the SKIN model developed by Gilbert et al. (2007), Pyka et al. (2007) and Ahrweiler et al. (2011), and it extends further the recent work of Blom et al. (2014). The chapter applies the SKIN model to a new context—the defence industry—and in doing so it introduces some novel features in the SKIN model framework. The following table points out the new aspects that are introduced in this model. All other parts of the model that are not discussed in this table are based on the original version of the SKIN model

attempt to reproduce the peculiarities and idiosyncrasies that characterize the defence industry. Table 6.1 points out specifically the new aspects that are introduced in the present chapter to extend the original SKIN model.[1]

6.3.1 Agents

Figure 6.1 presents a diagram describing the behavior of agents (private firms) and their market interactions within any given period t. Defence firms (business companies producing defence material, equipment and products) are the micro agents in the model. Agents differ from each other in two main respects (*firm heterogeneity*): financial capital endowment and technological capabilities (Gilbert et al. 2007; Pyka et al. 2007; Ahrweiler et al. 2011). First, we assume that defence companies have different endowments of financial capital, which they use both for their productive and innovative activities. The financial capital available to an enterprise i at time t is denoted by FC_i^t. Further, a fraction a of the population of firms is represented by large enterprises, whose financial capital is $b\mathrm{FC}_i^t$, where b is

[1] The model presented in this chapter is based on, and extends further, the recent work of Blom et al. (2014). The latter presents a more extensive presentation of the model's properties and focuses on its implications in terms of export dynamics, whereas the present chapter shifts the focus to the analysis of the concentration of the defence industry and how this evolves over time subject to a process of policy change and liberalization.

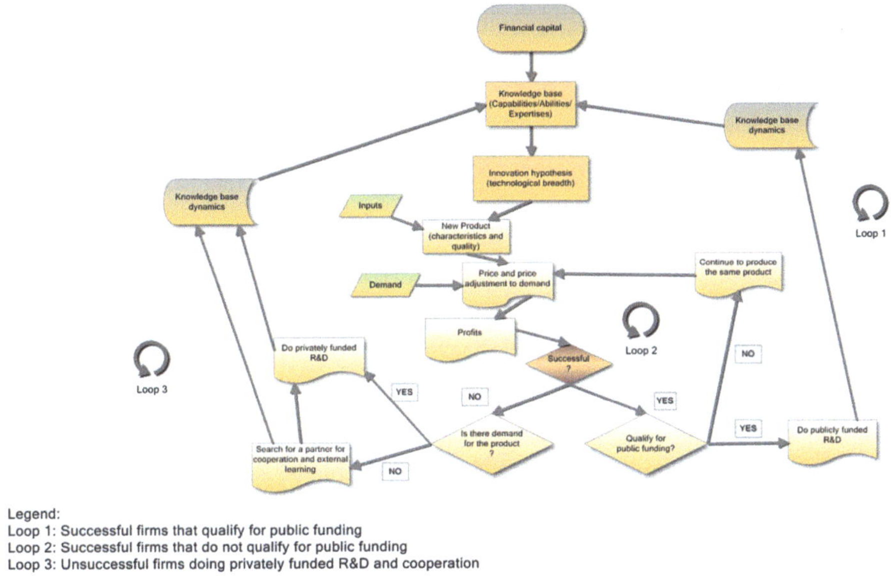

Fig. 6.1 Model flowchart

a positive constant ($b > 1$). This representation is in line with the firm size distribution in the defence industry, where a relatively small group of large firms co-exist with a larger number of SMEs.

Secondly, defence firms differ in terms of their knowledge base, i.e. the pool of scientific and technological competencies and skills that they employ in their innovative activities.[2] Following the SKIN model, we represent the firm's knowledge base as a set of units of knowledge. Each unit is a vector composed of three elements (Gilbert et al. 2007; Pyka et al. 2007; Ahrweiler et al. 2011):

- The *capability* (C), which defines a domain or area within the defence industry (e.g. weapon production). It is represented in the model as a randomly chosen integer between 1 and 1,000.
- The *ability* (A), defining a specific ability or skill that the firm possesses in this C domain (e.g. aerodynamic design technologies for platforms and weapons). It is a randomly chosen integer between 1 and 10.
- The *expertise* (E), which indicates the level of expertise that the firm has in using the ability A. This is also represented as a randomly chosen number between 1 (lowest) and 10 (highest).

[2] In the original version of the SKIN model, the knowledge base is labelled *kene*.

The knowledge base of firm i at time t is hence denoted by the matrix:

$$\mathrm{KB}_i^t = \left[\mathrm{C}_{ij}^t; \mathrm{A}_{ij}^t; \mathrm{E}_{ij}^t\right], \quad \text{for } j = 1 \ldots \mathrm{KL}_i^t \tag{6.1}$$

where KL_i^t is the length of the knowledge base of the firm, which indicates the breadth of its technological and competence portfolio. This length is defined as:

$$\mathrm{KL}_i^t = \ln\left[1 + \mathrm{FC}_i^t/c\right] \tag{6.2}$$

where c is the *capital-knowledge ratio*. This ratio is a positive constant ($c > 1$), which is assumed to be the same for all firms in the industry, representing the idea that the technological content (or regime) of each sector defines the amount of financial resources that it is necessary to invest in order to obtain a given amount of units of knowledge (this is analogous to the concept of *level of technological opportunities* of an industry, that is often used in the evolutionary economics literature on industrial dynamics). Given the representation in Eq. (6.2), the length of the knowledge base of the firm (KL_i^t) is higher for lower values of c (i.e. for knowledge-intensive industries), and it is positively related to FC_i^t. This means that the larger the financial capital available to the enterprise at a given time t, the greater the breadth of its technological and competence portfolio, and hence the pool of knowledge resources that it can use in its innovative activities. Equation (6.2) hence introduces an explicit link between financial capital endowments and the knowledge base of a company.

Defence firms compete in a highly innovative and technologically sophisticated environment. The model assumes that all firms in the market actively use their knowledge base in the attempt to create new products and processes. Innovative activities are represented in the SKIN model in such a way that, at any period t, each company formulates an *innovation hypothesis*, i.e. an idea or a plan for developing a new product or process (Gilbert et al. 2007; Pyka et al. 2007; Ahrweiler et al. 2011). The model represents this innovation hypothesis (IH_i^t) as a subset of the firm's knowledge base: the enterprise focuses on a specific subset of its technological competence (capabilities, abilities, expertise) that it finds particularly promising and worth developing further. The subset of expertises E_{ij}^t used in the innovation hypothesis is assumed to increase by one unit in the period, whereas those that are not used decrease by one unit. This represents *learning by doing* and *forgetting* mechanisms (Gilbert et al. 2007; Pyka et al. 2007; Ahrweiler et al. 2011).

The length of the innovation hypothesis, i.e. the number of units of knowledge that the company combines together in its innovative activities, defines the firm's *competence breadth* (CB_i^t). This is a positive integer whose minimum value is 2 and maximum value is KL_i^t (the overall length of the firm's knowledge base). The competence breadth measures how broad (or narrow) the technological competence of the firm is at any given period t, only focusing on the set of capabilities, abilities and expertise that the enterprise effectively uses in its innovative activities in the period (and disregarding all other units of knowledge that it does not utilize). The competence breadth CB_i^t of the enterprise is greater the greater the length of its knowledge base and, given the expression in Eq. (6.2), it is therefore dependent on the amount of financial capital available to the company. In other words, larger firms will on average have a larger competence breadth than SMEs.

6.3.2 Economic Environment

In any period t, each enterprise uses its innovation hypothesis to try to develop a new product. The outcomes of the innovative process are subject to a high degree of uncertainty and introduce an important stochastic element in the model. The new product is characterized as an index number that depends on the number of capabilities and abilities entailed in the innovation hypothesis according to the function:

$$\mathrm{PROD}_i^t = \left[\sum\nolimits_j \mathrm{C}_{ij}^t \cdot \mathrm{A}_{ij}^t\right] \mathrm{modN} \tag{6.3}$$

where N is the maximum number of different products (Gilbert et al. 2007; Pyka et al. 2007; Ahrweiler et al. 2011). The product is therefore characterized by the *breadth* of the innovation hypothesis, i.e. the number of different capabilities and abilities that the firm masters and it is able to combine in the development of the new artifact. The *quality of the product* is instead defined as:

$$\mathrm{PQ}_i^t = \left[\sum\nolimits_j \mathrm{A}_{ij}^t \cdot \mathrm{E}_{ij}^t\right] / \mathrm{CB}_i^t \tag{6.4}$$

Product quality is then obtained by multiplying the abilities and expertise levels for each of the vectors composing the innovation hypothesis and then normalizing the result by the competence breadth of the firm. This formulation implies that product quality depends on the *depth* of the innovation hypothesis (i.e. the set of abilities and expertise levels that are used in innovative activities), but it does not depend on its *breadth*. Therefore, SMEs, despite their lower financial capital endowment and more narrow knowledge base than large enterprises, may well be able to produce high-quality products, if they master a set of relevant abilities in a specific market segment. As explained below, this trade-off between competence breadth *versus* depth is an important characteristic driving the model's outcomes.

Given the quality of the new product, the firm sets its price as:

$$\mathrm{PR}_i^t = \mathrm{PR}_F^t \left[\exp\left(\mathrm{PQ}_i^t\right) - 1\right] \tag{6.5}$$

where PR_F^t is the final product price that was set at the beginning of the period. According to Eq. (6.5), the price of the new product PROD_i^t is a function of its quality PQ_i^t.

In order to produce the new product, the firm searches for inputs (e.g. capital equipment) in the market. The type of inputs it needs depends on the characteristics of the new product it is developing ($PROD_i^t$): the latter defines the set of inputs PK_{ij}^t that the firm must purchase, which are sold by intermediate producers at the price PR_{ij}^t. Intermediate producers that commercialize capital equipment and raw materials represent a share d of the population of firms in the market, whereas the corresponding percentage of end-products enterprises is denoted by e (hence: $e + d = 1$). Each end-product firm searches for the inputs it needs among the intermediate producers in the market, and it eventually purchases those with the lowest price and, *ceteris paribus*, the highest quality. If the enterprise does not find any input in the market at a price it can afford, it will not enter the production process.

Once the new product is ready for market commercialization, the firm sells Q_i^t units of it. The model assumes that demand patterns differ for different segments of the defence industry. On the one hand, intermediate products are sold to other firms within the defence sector only if there is a sufficient number of customers that want to purchase the capital good. On the other hand, new products that are destined to end users are always absorbed by the market. This assumption is in line with the fact that, in the defence industry, public procurement assumes a pivotal role, i.e. defence authorities typically purchase a substantial amount of new (or existing) defence products and material from domestic firms in order to secure military capabilities and so achieve national defence strategic objectives.

Given these market interactions and outcomes, at any period t the firm achieves a certain level of profit—which is largely dependent on the characteristics and quality of the new product it sells. Firm's i profit at time t is given by the difference between its revenues and its production costs:

$$\Pi_i^t = \left[PR_i^t Q_i^t\right] - \left[\sum\nolimits_j PR_{ij}^t \cdot PK_{ij}^t\right] \tag{6.6}$$

In order to represent the export activities of the firm—an aspect that was not considered in the original version of the SKIN model—we further assume that if the enterprise's revenues are large enough to cover sunk export costs, then the enterprise is able to extend the commercialization of its new product to foreign markets. This is in line with the key idea of the recent literature on firm heterogeneity and international trade (e.g. Melitz 2003; Helpman et al. 2004; Castellacci 2011, 2014), according to which only the most successful and productive enterprises within each sector are able to pay sunk export costs and overcome trade barriers in international markets, whereas most other companies will only produce for the domestic market. As illustrated in Melitz (2003) seminal model, in each industry there exists a threshold level of profits, EXP_{MIN}, that makes it possible to afford the sunk costs of internationalization and that only the most successful firms can achieve. Hence, if $\Pi_i^t < EXP_{MIN}$, an enterprise will not have sufficient revenues to invest in international activities, and will only produce for the domestic market; by contrast, if $\Pi_i^t \geq EXP_{MIN}$, the firm will also be able to export.

In this case, the additional profits accruing from the export of the new product are denoted by:

$$\Pi_{i\,EXP}^{t} = f\,\Pi_i^t - SC_i^t \tag{6.7}$$

where SC_i^t is the fixed (sunk) costs incurred by exporting firms, and f is the *export intensity*, i.e. the share of the firm's turnover that is commercialized in foreign markets ($0 < f < 1$).[3]

6.3.3 Performance Adjustment and Feedback Loops

At the end of the period t, having concluded its production, domestic sales and export activities, the firm is now endowed with an amount of financial capital given by the sum of the capital at the beginning of the period, the profits obtained from domestic sales and the additional profits from export activities:

$$FC_i^{t+1} = FC_i^t + \Pi_i^t + \Pi_{i\,EXP}^{t} \tag{6.8}$$

The firm does also adjust the price of its product. Depending on the market demand that the company has met for the new product $PROD_i^t$, an adjustment mechanism increases (decreases) its price if the demand level has been high (low). Specifically, the adjusted price for the product is:

$$PR_t^{t+1} = PR_i^t + gPR_i^t \tag{6.9}$$

where g denotes the percentage price change, which is positive if $Q_i^t > 0$, and negative if $Q_i^t = 0$.

Then the firm reassesses its innovative activities and strategies. It looks at its current market performance (i.e. the profits it has realized at time t), and decides whether this is satisfactory or not, and how it can be improved in the future. If $[\Pi_i^t + \Pi_{iEXP}^t] \geq \Pi_{MIN}$, then the enterprise is satisfied about its profit level; if, by contrast, $[\Pi_i^t + \Pi_{iEXP}^t] < \Pi_{MIN}$, the company is not satisfied and will try to improve its performance by adjusting its technological trajectory and strategy. The parameter Π_{MIN} denotes the *success threshold*, which indicates the profit level that marks the distinction between successful *versus* unsuccessful

[3] For simplicity we assume that all exporting firms sell abroad a given fraction f of their producton, and that this fraction f (export intensity) is a parameter that is exogenously defined for all firms belonging to the same market (sector, country). In line with our *stylized fact 3*, this parameter is arguably higher for firms in the defence industry than in other sectors of the economy. Further, in our simulation exercises, we will set a higher export intensity for enterprises in small open economies than in larger countries, since the former do on average sell a higher share of their turnover in foreign markets than the latter.

performance. This parameter is exogenously set at a given level for all firms in the market (sector, country). For simplicity, we start by assuming that the success threshold corresponds to the mean profit level in the industry in t_1: enterprises whose profits are above (below) the industry-level mean will be satisfied (not satisfied) with their current performance. There are two ways in which an enterprise can improve its performance over time. One can only be pursued by successful innovators, whereas the other is followed by unsuccessful performers.

New publicly-funded R&D projects If a firm is satisfied with the profits it has realized ($[\Pi_i^t + \Pi_{iEXP}^t] \geq \Pi_{MIN}$), it will apply for public funding for improving its product further (publicly-funded incremental innovation). This is a new feature that was not present in the original version of the SKIN model. It is a very relevant aspect in the context of the defence industry, since public defence authorities typically finance a substantial amount of domestic R&D activities through public procurement. We assume that this public funding is granted to the applicant according to two complementary criteria: (1) The quality of the firm's product has to be above a given *product quality threshold* ($PQ_i^t \geq PQ_{MIN}$); (2) The firm's competence breadth has to be above a given *competence breadth threshold* ($CB_i^t \geq CB_{MIN}$). The intuition behind this public funding allocation mechanism is in line with the practice that public defence authorities typically follow. To put it simply, when policy-makers evaluate the possibility to finance a company to develop a new product, they look at both: (1) the quality of its current product, which gives them an indication of the likelihood that the firm will be able to produce a successful incremental innovation of it in the future; and (2) the breadth of the firm's expertise in several different areas, since large multi-product and multi-competence enterprises are those that have presumably received public funding already in the past and thus previously developed a reliable user-producer relationship with public authorities.

If an enterprise meets both requirements, it will get access to additional resources to finance its innovation activities. Its financial capital will then further increase by a lump sum PS (the *public subsidy* granted by national authorities). This will make it possible to enlarge the firm's knowledge base [see Eq. (6.2)] and hence formulate a new innovation hypothesis with a higher competence breadth. If this innovation project is successful, the company will commercialize a new product $PROD_i^{t+1}$, which further improves its previous innovation $PROD_i^t$. However, if the attempt to introduce this incremental innovation is not successful, the enterprise will continue to produce and sell $PROD_i^t$. All in all, the possibility for successful firms to apply for new publicly-funded R&D projects introduces a source of cumulative causation in the model, since in any period *t* there will be a limited number of successful companies that will be able to qualify for public R&D support, which is likely to lead to further incremental innovation and satisfactory profits for them in the future.

Changing search strategies If a firm is not satisfied with the profits it has realized ($[\Pi_i^t + \Pi_{iEXP}^t] < \Pi_{MIN}$), it will try to improve its performance by starting to search in a new direction (Nelson and Winter 1982). An enterprise can apply two different search strategies to adjust its performance (Gilbert et al. 2007; Pyka et al. 2007; Ahrweiler et al. 2011).

1. **Privately-funded R&D projects**. If its product $PROD_i^t$ was sold in the market but the demand level was not sufficient to realize a satisfactory profit, the company will undertake a new R&D project funded through its own internal resources (as long as its financial capital endowment is sufficient to finance this new project). The new R&D project is of incremental nature: it will aim at improving one of the abilities (A_{ij}^{t+1}) in its innovation hypothesis, i.e. to achieve a better specialization and technological sophistication in the technological space on which it is currently focusing—hence to increase its technological *depth*, given its *competence breadth*. If the incremental innovation is successful, the firm will increase the expertise level E_{ij}^{t+1} related to the ability A_{ij}^{t+1} that the R&D project focuses on.
2. **Cooperation**. If the firm's product did not meet any demand in the market (hence leading to negative profits), this gives a clear indication that the firm's current innovation hypothesis is not well suited to the user requirements, and that it must be adjusted. The enterprise can do this by searching for an external partner for cooperation. The firm will first search among its previous partners, then its suppliers and users, by looking at the capability set they possess as indicated by their respective innovation hypotheses and market product characteristics. The probability that a company finds a suitable cooperation partner depends on the parameter COOP, which defines the *cooperation propensity* of firms in the market. When a firm finds a collaboration partner, it can add the partner's innovation hypothesis triples to its own, thus achieving a broadening up of its capability set and knowledge base. To put it differently, cooperation enables the exchange of knowledge among different agents, and this is likely to improve the performance of these by augmenting their respective knowledge bases and technological competencies.

   All in all, the two strategies pursued by unsuccessful performers—privately-funded R&D and cooperation—introduce a catch up mechanism in the model, since firms lagging behind the technological frontier may improve their technological position and adjust their market performance by means of such R&D and imitation strategies.

   A summary and overview of the model (see Fig. 6.1) highlights the following two key features of this theoretical framework. First, in any period *t*, the agents will be sorted in three distinct groups: (1) successful innovators that qualify for public funding (see Fig. 6.1, *loop 1*); (2) successful innovators that do not meet the criteria for public support (*loop 2*); (3) unsuccessful performers, which will either undertake a new privately-funded R&D project or imitate by searching a cooperation partner (*loop 3*). Secondly, the overall dynamics of the model, as shown in the next section, depends on the combination of two different mechanisms: (1) a cumulative causation mechanism according to which the best performers will tend to get public support and hence strengthen their market position even further in the future; (2) a catch up mechanism through which less successful companies will be able to adjust their performance and possibly achieve a leading market position in the future. Section 6.5 will analyze how these mechanisms shape the long-run properties of the model, and Sect. 6.6 will then investigate how future policy changes towards liberalization may affect export dynamics and market opportunities in the defence industry.

6.4 Variables and Indicators

6.4.1 Key Aggregate (Industry-Level) Outcomes

The following two variables are aggregate outcomes of the model, i.e. emergent properties that are observed at the industry-level as the result of micro-level behavior and agents' interactions. They represent the key variables defining the concentration of the defence industry, and thus the main factors we seek to explain in our simulation analysis.

Concentration index We use the C5 concentration index, defined as the total financial capital owned by the five largest firms in the market as a share of the total financial capital in the defence industry.

Export concentration ratio (%) We define this as the E5 export concentration index, i.e. the total value of export obtained by the five largest firms in the market as a percentage of the total value of export in the defence industry. Compared to the C5 index, this is a more specific variable measuring concentration patterns in relation to firms' export activities, which is an important aspect to look at when analyzing the effects of market liberalization.

6.4.2 Key Parameters: Environmental and Policy Characteristics

The following four parameters describe some key characteristics of the economic environment in which agents operate, which may be affected by policy actions and strategies over time. They represent the main explanatory variables of interest in our simulation analysis.[4]

[4] All other model's parameters not defined here are listed in Table 6.2 in Sect. 6.5.

Cooperation propensity (COOP) This defines the extent to which agents are willing (and able) to cooperate with others in the same market, i.e. their collaboration propensity. This parameter ranges on a continuous scale from 0.50 (lowest cooperation propensity) to 0 (highest propensity).

Success threshold (Π_{MIN}) This indicates the threshold above (below) which firms consider themselves satisfied (not satisfied) with their current market performance (e.g. corresponding to the mean profit level in the industry at t_I). The parameter is defined in the profit space ranging from 0 to 12,000. This parameter is largely dependent on the extent and intensity of market competition, i.e. the success threshold is higher (lower) in a more (less) open and competitive market, because agents must compete with a greater (smaller) number of competitors (including foreign firms) in order to maintain their market position. In other words, in a more (less) open and competitive market companies tend to be more (less) demanding because they are aware they face a stronger (weaker) competition.

Public funding requirement I: Product quality threshold (PQ_{MIN}) This is the first of the two criteria set by public authorities to grant public support to private defence firms. It ranges on a continuous scale defined on the quality domain between 0 (loose quality requirement, easy to get public funding) to 10 (strict quality requirement, difficult to get public support).

Public funding requirement II: Competence breadth threshold (CB_{MIN}) This is the second requirement for qualifying for public support. The parameter ranges on a continuous scale defined on the innovation hypothesis domain between 0 (narrow technological competence, easy to get public funding for most firms) to 10 (broad technological competence, difficult to get public support for many narrowly specialized companies).

6.4.3 Other Model Parameters

These parameters do also represent environmental characteristics affecting the industry dynamics. However, they may not be influenced by policy actions in the short-run.

Number of firms: Total number of enterprises in the market.
Number of products: Total number of products that are sold in the market.
Share of large firms: Number of large enterprises as a percentage of the total number of firms in the market.
Share of end products firms: Number of enterprises that produce final products as a percentage of the total number of firms in the industry.

6.5 Simulation of Different Policy Scenarios for EU Countries

Figure 6.2 depicts the time path of the two main outcome variables of the model—industry concentration index C5 and export concentration ratio E5—for a 500-t period (average of 50 replications). The graphs illustrate the basic working of the model. Both concentration indexes rapidly decrease during the first 100-t period, and then they remain rather stable thereafter. The export concentration ratio does rapidly stabilize at a level below 9 %.

Let us now use the model to understand this dynamics, and carry out a comparison of six different policy scenarios. The exercise is intended to compare the current situation—in which the European defence industry is characterized by a high level of national protection and a low degree of market liberalization—with five possible future scenarios, which will be realized when public defence authorities in European countries will implement the new EU Directive and thus introduce a stronger degree of openness and liberalization in this market.

Table 6.2 lists all parameters of the model and reports the values that we have assigned to them in the current scenario version of the model. The specific values that we have used to calibrate the parameters are purely indicative and do not correspond to real data for the defence industry in European countries (which are not available). The idea is to set up a stylized and simple comparison between a large and a small national defence market, and see whether and the extent to which these country-specific differences affect the outcomes of the model. Specifically, we present two versions of our calibration exercise, one for a small country (e.g. Sweden, Netherlands, Norway) and the other for a large economy (e.g. France, Germany, UK). The small country version has 150 defence enterprises, 60 products and an average export intensity of 50 % (i.e. we reasonably assume that in a small domestic market exporting firms do on average sell a substantial share of their defence products abroad). The large country version has instead 500 enterprises, 400 products and a 10 % mean export intensity (i.e. if the domestic market is large, exporting firms sell on average a greater share of their products at home and a smaller share abroad).

Despite the fact that this parameter setting does not intend to reproduce real empirical patterns of the EU defence industry in an accurate quantitative way, the resulting model dynamics does however resemble well some of the main stylized facts characterizing the industry. In this current scenario, the model describes an industry with large firm heterogeneity, stable and concentrated structure, high innovativeness, high export intensity and active public involvement.

After setting up the current scenario, we have then constructed five future scenarios that represent different possible trajectories that the industry may follow in the future as a result of different policy strategies in terms of the implementation of the new EU Directive. These five scenarios differ in terms of how rapidly and actively defence authorities of national Member States will decide to implement the new Directive and introduce market liberalization, i.e. the typology goes from a softer and more gradual implementation towards a more rapid and radical market reform.

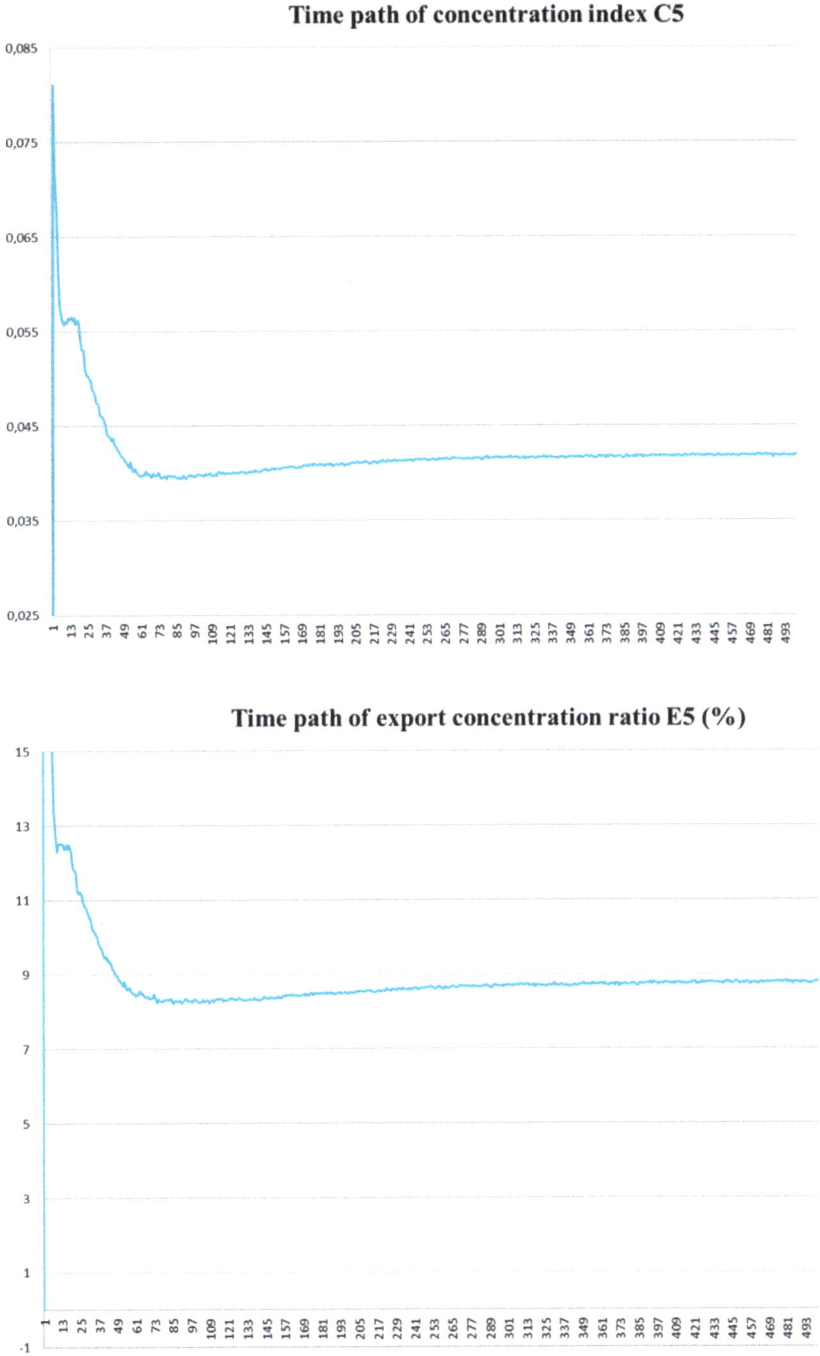

Fig. 6.2 Time path of the main outcome variables of the model (t = 1–500). Each point is the average of 50 replications

Table 6.2 Model's parameters and calibration values in the *current scenario*

Parameter	Definition	Large country	Small country
COOP	Cooperation propensity	0.45	0.45
Π_{MIN}	Success threshold	4,000	5,000
PQ_{MIN}	Product quality threshold	6	6
CB_{MIN}	Competence breadth threshold	6	6
M	Number of firms in the market	500	150
N	Number of products in the market	400	60
FC_i^0	Initial allocation of financial capital	10,000	10,000
a	Share of large firms	0.05	0.05
b	Large firms' multiplicative factor	10	10
c	Capital-knowledge ratio	20	20
d	Percentage of intermediate products firms	0.40	0.40
e	Percentage of end-products firms	0.60	0.60
EXP_{MIN}	Export threshold	6,640	6,640
f	Export intensity	0.10	0.50
g	Price adjustment share	0.10	0.10
PS	Public subsidy	10,000	10,000

1. *Higher success threshold scenario*: This represents a situation in which national policy-makers of, say, country X do not introduce any significant and active reform towards liberalization. However, the increased openness of other EU countries' defence sectors naturally induces a stronger degree of competition in country X's domestic market. Faced with the challenge posed by the entry of other European companies into the domestic market, country X's enterprises will react by adjusting their success threshold upward.
2. *Higher cooperation scenario*: If national defence authorities introduce measures aimed at promoting inter-firm collaborations (within and across countries), defence firms will tend to increase their cooperation propensity and, hence, their ability to exploit knowledge spillovers effects.
3. *Higher product quality threshold scenario*: Policy-makers may also decide to change public procurement mechanisms and modify the criteria they use to allocate public R&D funds to defence companies. This may be a natural consequence of the fact that foreign EU enterprises will be allowed to participate in national public procurement tenders, thus making these much more competitive and demanding for domestic firms. In particular, if national authorities decide to emphasize the first allocation criterion, they will increase the product quality threshold that companies have to satisfy in order to qualify for public support.
4. *Higher competence breadth threshold scenario*: By contrast, if they decide to focus on the second allocation criterion, they will increase the competence breadth threshold, and hence start to allocate more funds to large multi-product and multi-competence firms and fewer resources to smaller specialized suppliers.

5. *Market liberalization scenario*: Finally, if all the policy strategies indicated by the previous four scenarios are combined and implemented together, we obtain a full market liberalization scenario. This may be thought of as the most rapid and most radical way of introducing market liberalization in the defence industry.

Figure 6.3 reports the results of the policy simulation analysis. The graphs in this figure illustrate the dynamics of the concentration index C5 and the export concentration ratio E5 for a 300-t period (average of 100 replications), for both the large- and small-country versions of the model.

Looking at the current scenario in Fig. 6.3, this illustrates again the basic working of the model. Over time, firms tend to learn and improve their technological performance by means of learning by doing, cooperation and R&D activities, so that the number of exporters in the industry does gradually increase as time goes by. Correspondingly, the mean product quality in the industry will increase over time, the number of firms receiving public funding will increase, and the concentration level will therefore decrease. However, a comparison between the current scenario and the other five shows that the former is the one characterized by the worst performance in the long-run (i.e. the highest levels of C5, and the highest export concentration ratio at $t = 300$). Specifically, we observe the following five patterns.

1. In the *higher success threshold scenario*, companies are on average more responsive to market opportunities and more actively investing in product quality and technology upgrading *vis-à-vis* what they tend to do in the current scenario. This is also the scenario where firms undertake the greatest number of new privately-funded R&D projects in order to adjust and improve their technological performance. In the long-run, this leads to an increase in the number of well-performing companies, and hence a sharp reduction in the industry and export concentration ratios. A comparison of the small- and large-country versions of the model indicates that the main difference is in terms of the export concentration index. The decrease in the E5 concentration index over time (in percentage terms) is stronger in a large country than in a small economy.
2. The *higher cooperation scenario* is the one where defence firms, due to their higher collaboration propensity, exploit more actively the opportunities provided by external learning and knowledge spillovers effects. This explains why this scenario outperforms the current one leading to a much lower concentration ratio. Due to the strength of this imitation-based catch up mechanism, a substantial number of firms are able to attract public funding for carrying out R&D activities. Many of these SMEs and catching up enterprises are also able to start to export their products abroad, so that the E5 export concentration ratio does also decrease significantly over time (for the large country the decrease is more pronounced than for the small-country).
3. The *higher product quality threshold scenario* also leads to a better export performance in the long-run and a lower market and export concentration. Differently from the previous, though, in this policy scenario enterprises are able to strengthen their market position over time despite the fact that public

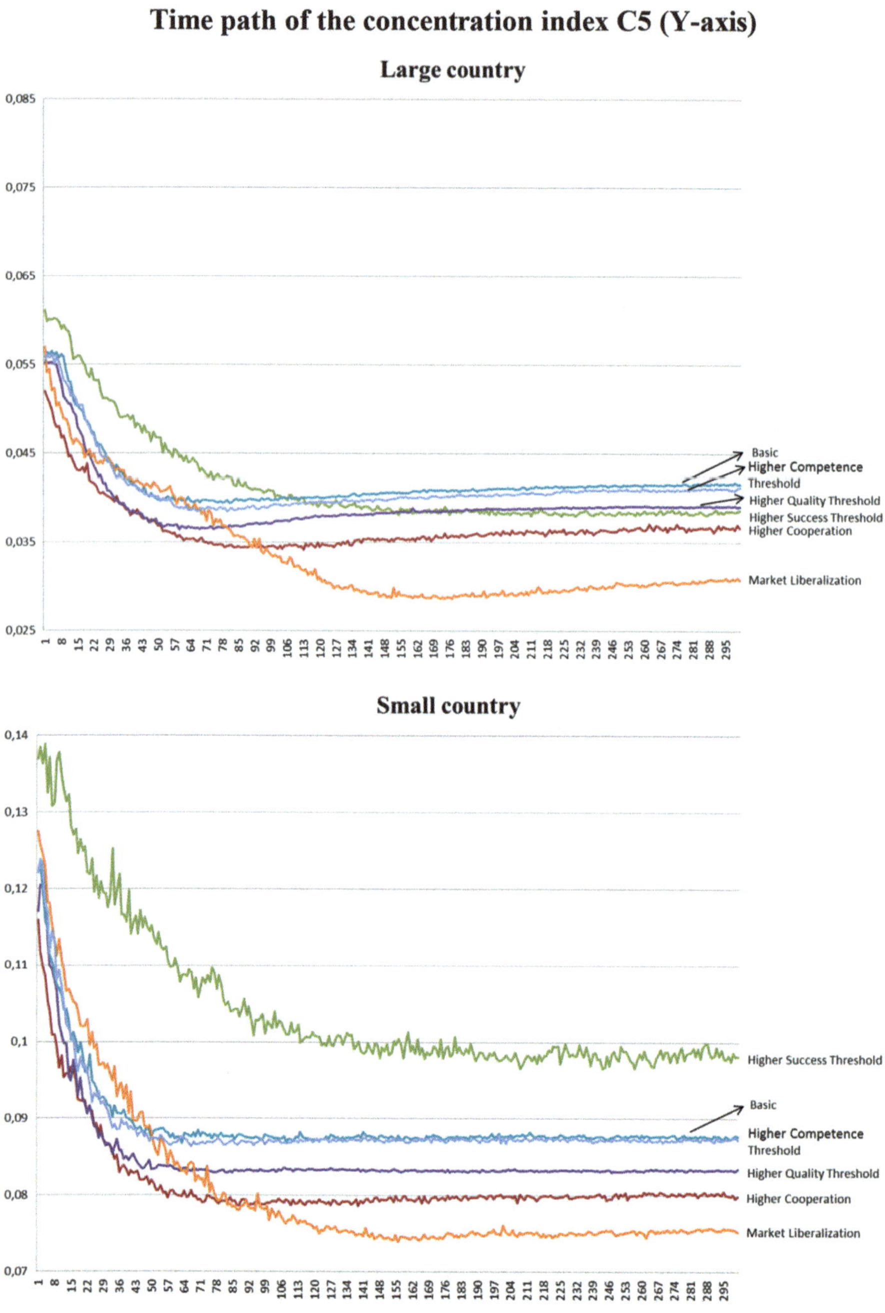

Fig. 6.3 (continued)

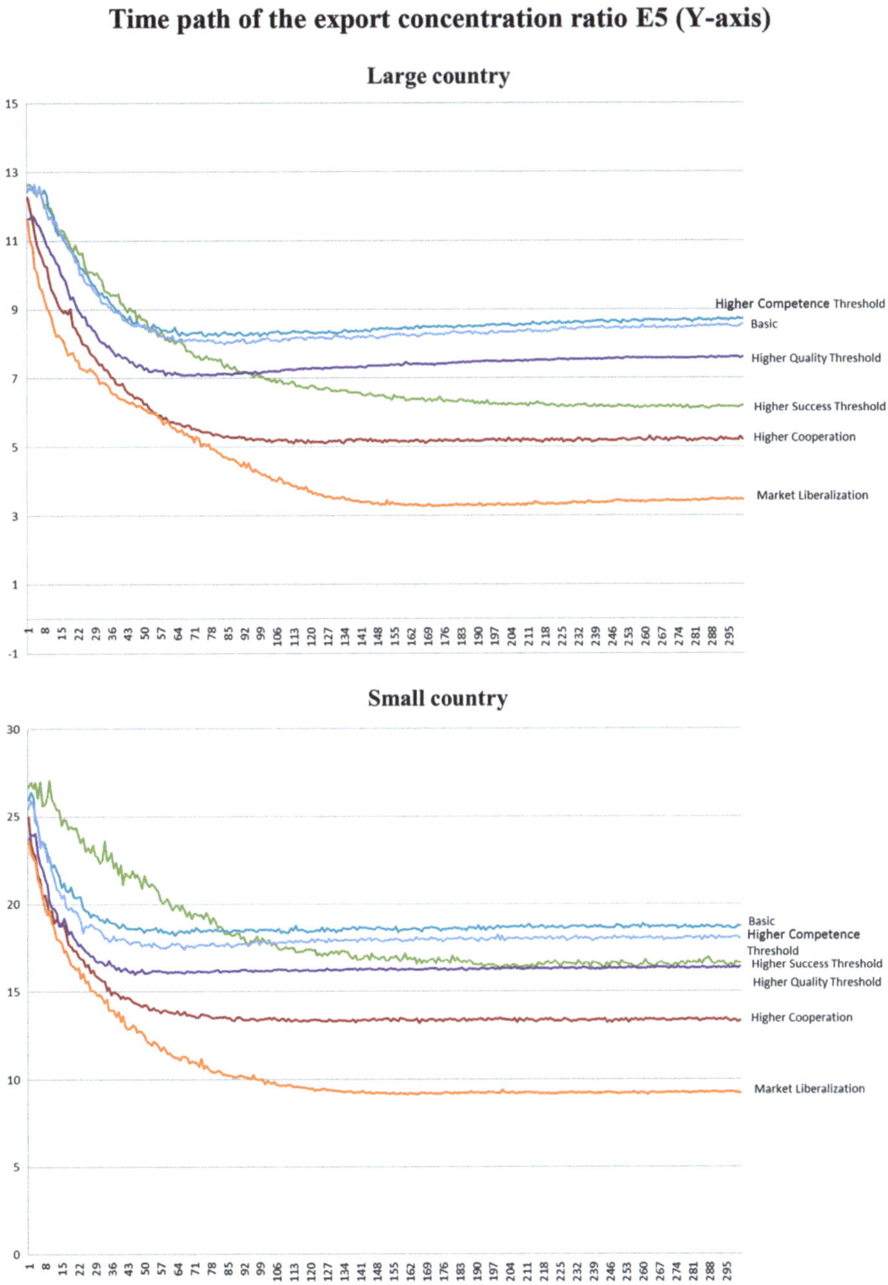

Fig. 6.3 Simulating six different policy scenarios (each point is the average of 100 replications)

funding opportunities decrease. The decrease in export concentration is again relatively stronger in the large- than the small-country version of the model. The reason is that the competition and selection mechanisms triggered by product quality enhancing investments are magnified and arguably have stronger effects in a large market than in a small economy.

4. The *higher competence breadth threshold scenario* does instead not lead to any substantial change as compared to the current scenario. This is because this second mechanism of public funds allocation (competence breadth) is not aligned with the crucial market requirement for competing in international markets (technological depth). Hence, this will end up by strengthening the leading position of large oligopolistic producers but will not increase export and market opportunities for most other SMEs in the market.
5. Finally, the *market liberalization scenario* clearly outperforms all other policy strategies considered in Fig. 6.3, since this is obtained by combining together all four previous scenarios, representing the possibility that national defence authorities will opt for a rapid and radical reform of the defence market towards openness and full liberalization. This would lead, according to this model, to a substantial increase in the number of exporting firms and a more competitive and less concentrated (domestic and export) market in the long-run. The effects of full market liberalization on industry concentration are again relatively stronger for a large country. Interestingly, however, these effects are more rapid in the small-country version of the model, whereas in the large-economy version the market liberalization scenario takes a substantially longer time (between 50 and 100 periods) before overtaking the others.

6.6 Conclusions

This chapter set out to investigate defence industry concentration in the context of a liberalizing EU market. It aimed to study (1) to what extent a more open market and reduced policy protection will affect industry concentration in the European defence industries, and (2) whether there are important differences in the ways that the defence sectors of larger and smaller European countries are affected.

In terms of the effect of a more open market and reduced policy protection on industry concentration, this chapter analysed changes along four parameters that represent direct consequences of the liberalization process—increases in firms' collaboration propensity, greater degree of competition they face in the domestic market, and changes in the criteria for allocation of R&D policy support, with a consequent reduction in the amount of policy support that defence based companies receive. The first two scenarios represent a more cautious and passive approach from the governments—where they simply relaxed the most visible protectionist policies such as discriminatory procurement practices and offset requirements, and thereby increase competition from foreign firms and improve opportunities for collaboration with foreign suppliers. The second two scenarios represent instead a more active strategy of implementation of the EU Directive—where national governments increase the threshold for receiving R&D funding based on their importance for economic growth and/or national security objectives, and thereby force national companies to either improve the quality of their innovative products or increase their competence breadth to qualify for continued funding.

The simulation analysis found that all four liberalization parameters have a negative effect on industry concentration, i.e. the liberalization process is likely to make defence industries in European countries less concentrated than they are at present. Further, we also found that a more radical and rapid implementation of the liberalization reform (obtained in the model by combining changes in all of the liberalization parameters) has a larger effect on industry concentration than the other more cautious and gradual policy strategies we have considered. The main reason explaining the effects of market liberalization on industry concentration is that defence companies—although facing in the short-run substantial challenges due to the higher degree of competition they experience in their domestic market—are in the SKIN model able to improve their dynamic performance by collaborating with other companies and investing in new (public and private) R&D projects.

In terms of differences in the ways the defence industries of larger and smaller European countries are affected, this chapter found that the larger countries experience a relatively greater reduction in industry concentration than the smaller countries. The analysis was based on existing studies of the European defence industry and assumed that the smaller economies had a higher initial concentration levels than the larger countries, as it is typically the case. The analysis found that the higher number of companies in the larger countries' markets resulted in a more favourable dynamics that reduced the industry concentration relatively more than in the smaller countries. The reason for this, according to the model considered here, is that the competition, selection, and learning mechanisms triggered by the liberalization process are arguably stronger in a larger market.

Based on the analysis carried out in this chapter, we conclude that the European countries will in the long-run benefit from the introduction of the EU Directive. If the Directive will effectively manage to foster a more open and competitive European defence sector, industry concentration will decline both in the smaller and larger European countries. This will limit incumbent companies' possibilities to extract monopoly rents and allow the European countries to acquire defence equipment more cheaply. This would tend to change the industry structures in the European countries from stable oligopolies where incumbents mostly engage in incremental and cumulative types of innovation to more dynamic and competitive industries where innovators can also engage in radical and exploratory innovative activities.

Acknowledgment The chapter is produced as part of the INNDEF project. Financial support from the Norwegian Research Council (FORFI programme) is gratefully acknowledged. A previous draft of this paper was presented at the SKIN Model Workshop in Koblenz, June 2012. We wish to thank Petra Ahrweiler, Nigel Gilbert, and Andreas Pyka for the very helpful comments and suggestions.

References

Ahrweiler P, Gilbert N, Pyka A (2011) Agency and structure. A social simulation of knowledge-intensive industries. Comput Math Organ Theor 17:59–76

Blom M, Castellacci F, Fevolden A (2014) Defence firms facing liberalization: Innovation and export in an agent-based model of the defence industry. Comput Math Organ Theor, in press

Brauer J, Dunne P (2004) Arms trade and economic development: theory, policy and cases in arms trade offsets. Routledge, London

Castellacci F (2011) Technology, heterogeneity and international competitiveness: insights from the mainstream and evolutionary economics paradigms. In: Jovanovic M (ed) International handbook of economic integration. Edward Elgar, Northampton, MA

Castellacci F (2014) Service firms heterogeneity, international collaborations and export participation. J Ind Compet Trade 14(2):259–285

Castellacci F, Fevolden A (2014) Capable companies or changing markets? Explaining the export performance of firms in the defence industry. Defence and Peace Economics (in press)

Edwards J (2011) The EU Defence and Security Procurement Directive: a step towards affordability? International Security Programme Paper 2011/05. Chatham House, London

Gilbert N, Ahrweiler P, Pyka A (2007) Learning in innovation networks: some simulation experiments. Phys A 378:100–109

Helpman H, Melitz M, Yeaple S (2004) Export versus FDI with heterogeneous firms. Am Econ Rev 94(1):300–316

Klepper S (1996) Entry, exit, growth, and innovation over the product life cycle. Am Econ Rev 86(3):562–583

Kuschel J (2008) The European defence market, great challenges – small moves. Verlag, GRIN Verlag GmbH

Malerba F, Orsenigo L (1996) Schumpeterian patterns of innovation are technology-specific. Res Pol 25:451–478

Markowski S, Hall P, Wykie R (2010) Defence procurement and industry policy: a small country perspective. Routledge, London

Melitz M (2003) The impact of trade and intra-industry reallocations and aggregate industry productivity. Econometrica 71(6):1695–1725

Nelson RR, Winter SG (1982) An evolutionary theory of economic change. The Belknap Press of Harvard University Press, Cambridge, MA

Pyka A, Gilbert N, Ahrweiler P (2007) Simulating knowledge-generation and distribution processes in innovation collaborations and networks. Cybern Syst 38(7):667–693

Chapter 7
Testing Policy Options for Horizon 2020 with SKIN

Petra Ahrweiler, Michel Schilperoord, Andreas Pyka, and Nigel Gilbert

Abstract This chapter is about a SKIN application to the world of EU-funded research networks in the area of information and communication technologies (ICT). The application was commissioned by the DG Information Society and Media (DG INFSO) as an impact assessment of the funding strategies in the 7th Framework Programme (FP7) and ex-ante evaluation of the upcoming funding cycle called Horizon 2020. The focus of this chapter is on the changes of the SKIN model to become SKIN-INFSO, the strategy to calibrate the adapted SKIN model with empirical data from the European Commission to achieve realistic simulation results, and the ways we analysed and validated our results using network analysis. Details of the policy experiments using the SKIN-INFSO application for the study and their results are reported elsewhere [Ahrweiler, Gilbert, Pyka, Innovation policy modelling with SKIN. In: Johnston E et al (eds) Policy informatics. MIT Press, Cambridge, 2014, forthcoming].

P. Ahrweiler (✉)
EA European Academy of Technology and Innovation Assessment GmbH, Wilhelmstr. 56, 53474 Bad Neuenahr-Ahrweiler, Germany
e-mail: Petra.Ahrweiler@ea-aw.de

M. Schilperoord
UCD Innovation Research Unit (IRU), UCD Michael Smurfit Graduate Business School, University College Dublin, Carysfort Avenue, Blackrock, Dublin, Ireland
e-mail: michel.schilperoord@ucd.ie

A. Pyka
Universität Hohenheim, 70593 Stuttgart, Germany
e-mail: a.pyka@uni-hohenheim.de

N. Gilbert
Centre for Research in Social Simulation, University of Surrey, Guildford GU2 7XH, UK
e-mail: n.gilbert@surrey.ac.uk

N. Gilbert et al. (eds.), *Simulating Knowledge Dynamics in Innovation Networks*, Understanding Complex Systems, DOI 10.1007/978-3-662-43508-3_7,

7.1 Introduction

Recent studies of network architectures and their dynamics indicate that research and innovation performance may be related to specific network structures and mechanisms of network evolution. From this it follows that there is scope for the improvement of the policy instruments used to create and support research and innovation networks. The INFSO Evaluation Unit has already commissioned three studies using network analysis to address the structuring effects of Framework Programme Information and Communications Technology (ICT) networks, their international reach, and the linkages between EU research and deployment and regional innovation systems. Very important was the recommendation to continue to ensure that support for SMEs and for large firms is not "compartmentalised" into different measures or tools (European Commission/DG INFSO 2008: 3) so that these organisations can be fully integrated partners in sustainable EU-funded research networks. This would be necessary to increase the amount and the speed of knowledge transfer among organisations. Also interesting was the recommendation of the Evaluation Panel to develop a focus on the appropriateness of partnerships, in particular the inclusion of highly innovative participants (European Commission/DG INFSO 2008: 6). Moreover, it was reported that the scope and opportunity for evaluation insights provided by network analysis could be extended by using simulation methods (European Commission 2009: 15f).

The methodological framework we proposed in response had to cover the following tasks:

- Use the DG INFSO database on FP7 ICT networks as data input to calibrate the simulation model
- Process these data in simulation experiments with different policy scenarios
- Analyse and interpret the simulated network data for Horizon 2020 impact assessment/ex-ante evaluation

7.2 Adapting the SKIN Model to Become INFSO-SKIN

For these purposes, we built from a SKIN version that had already been adapted to the application context of FP-funded R&D networks in the EU project "Network Models, Governance and R&D Collaboration Networks" (NEMO, NEST, FP6; Scholz et al. 2010). We adapted the SKIN model to the study scope by developing new types of agent to represent research institutes, SMEs and big firms as required by the policies and scenarios to be modelled. The behaviours of the agents follow our empirical understanding of the processes of network formation and evolution in the Framework Programmes.

7.2.1 Agents and Kenes

In the INFSO-SKIN model, agents are R&D organisations (universities and research institutes), R&D departments in Large Diversified Firms (LDFs) and Small and Medium sized Enterprises (SMEs). Each agent has a kene, which in this case is a “quadruple”. A quadruple includes the following elements: **R**esearch **D**irection (RD) represented by an integer, **C**apabilities (C) represented by an integer, **A**bilities (A) represented by a real number and **E**xpertise (E) represented by an integer.

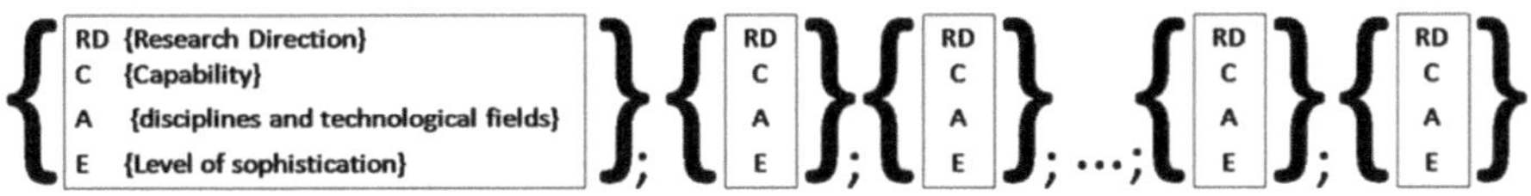

The research direction (RD) represents the general knowledge orientation, which can be related to basic or applied research. A value of 0 indicates completely theoretical knowledge (i.e. basic research); higher values indicate an increasingly applied orientation; a value of ‘9’ means exclusively applied research. For every knowledge field, applied as well as fundamental research directions exist. These research directions are attached to the actors’ quadruples to allow for a differentiation between e.g. university and firm actors. The capabilities (C) represent different knowledge areas and technological disciplines e.g. biochemistry, telecommunications or mechanical engineering. For modelling purposes the number of different capabilities has to be chosen to be large enough to cover all potential research areas that are encompassed in the European Framework Programmes. The knowledge space is structured: for example 800 different capabilities are divided equally among each of eight themes. The ability (A) represents the actors’ specialisation in the capabilities’ fields. In biochemistry, for instance, abilities are protein design, genomics, combinatorial chemistry, bioinformatics, filtering etc. The expertise (E) stands for the advancement of an actor’s skills in the respective knowledge field.

In INFSO-SKIN, we implemented three different types of agents: research institutions including universities (RES), large diversified firms (LDFs) and small and medium-sized enterprises (SMEs), see Table 7.1.

7.2.2 Behaviours and Settings

The general behaviours of the agents follow an empirical understanding of the phases in network formation and evolution of the Framework Programmes.

Table 7.1 Agent types and attributes

Agent types	Contribution (indicated by length of kene)	Objectives	Research direction	Capacity for partnerships
RES	Variety of knowledge	Publications, patents	Basic or applied	Large (>2)
LDFs	Variety of knowledge	Patents	Applied	Large (>2)
SMEs	Specialised knowledge	Patents, publications	Applied	Small (1 or 2)

7.2.2.1 Definition of Policy Incentives and Rules for R&D Collaboration

The EU provides funding for collaborative research. The rules are defined in the Framework Programmes (e.g. rules for project consortia, research topics, time span of the FP etc.). Actors (research institutes, firms etc.) want to apply for funding. The Calls of the Commission specify:

- The type of instrument (IP or STREP). This specifies the minimum number of partners in a consortium, the composition of partners, and the length of the project.
- The date of Call (to determine the deadline for submission)
- A range of capabilities, a sufficient number of which must appear in an eligible proposal (how many is sufficient depends on the type of instrument, e.g. lower for a narrowly focused STREP than for a IP with a broad scope)
- The funding available for this Call
- The number of projects that will be funded
- The desired basic or applied orientation

7.2.2.2 Process of Consortium Formation/Partner Choice

The actors form project consortia. Partner choice mechanisms apply. Firstly, the agent looks at the list of its previous partners. Secondly, previous partners that agreed to join the proposal can add previous partners from their list. Thirdly, new partners will be searched for. The search process is guided by the requirements outlined in the Call, a list of capabilities. The proposal is considered to be eligible only if a sufficient number of these capabilities appear. If no agent from the list of previous partners can contribute such a capability in the first iteration, then in the second iteration previous partners of those agents that agreed to join the proposal can ask their previous partners. If the required capability is still not found, the proposal consortium can search for the knowledge in the population of all actors. This is done on a random basis. In each iteration, n agents can be asked whether they have the respective capability and whether they want to join the proposal consortium. The possibilities to join a proposal consortium are determined by the same rules as for the determination of project initiations (see next step). The length of the kene determines whether the agent has free capacities for new activities. For example, an SME, whose kene is of minimum size (i.e. five quadruples) and which is already in a project or a proposal consortium has to reject the offer.

The agents in a collaboration produce a proposal, representing the relevant knowledge of the partners. The consortium submits the proposal to the Commission. Each agent contributes one or more capabilities. The upper bound on how many capabilities an agent can contribute depends on the size of the agent's kene. If the agent has only one of the capabilities specified in the call, it contributes this capability. If the agent has none of the required capabilities, it declines to join. A proposal will be submitted if a sufficient number of capabilities appear.

7.2.2.3 Process of Proposal Selection

The Commission evaluates the proposals according to a template that emphasises the contents (programme match), and the quality and architecture of the consortium (e.g. minimum number of members, industry involvement etc.). Proposals need to have sufficient partners with a sufficient number of capabilities as specified in the Call to be considered eligible. All proposals that fulfil the eligibility criteria are then ranked according to the average expertise level of the proposals (i.e. the expertise levels of the capabilities are summed and divided by the number of quadruples in the proposal). If some proposals turn out to have the same average expertise level, the tied proposals are ordered according to the number of the capabilities in the proposal that were specified in the Call. If, after the application of this rule, proposals are still ranked equally, one of them is selected at random. The number of proposals specified in the Call are then selected working down the ranked list. Proposal consortia that are not successful are dissolved. Those consortia that were selected start their projects.

7.2.2.4 Process of R&D Cooperation

The projects start to work on a "project hypothesis", i.e. they are involved in research and cooperative learning activities. They produce deliverables (e.g. a number of publications and patents). The research in the projects follows the ideas of SKEIN (Scholz et al. 2010). Agents in project consortia are randomly allocated to sub-projects and combine their kenes. Every 3 months they produce an output (deliverable), which can be a publication or a patent. A transformation function for the project hypothesis produces (1) a number between 0 and 1, which decides on the type of outcome and (2) a figure describing the probability of success of the project hypothesis. The potential outcome (publication or patent) depends (a) on the research orientation of actors (i.e. an applied research orientation increases the probability of a patent whereas a basic research orientation decreases this probability); (b) the variance in capabilities involved in a project hypothesis—the lower the variance, the lower the probability of a patent.

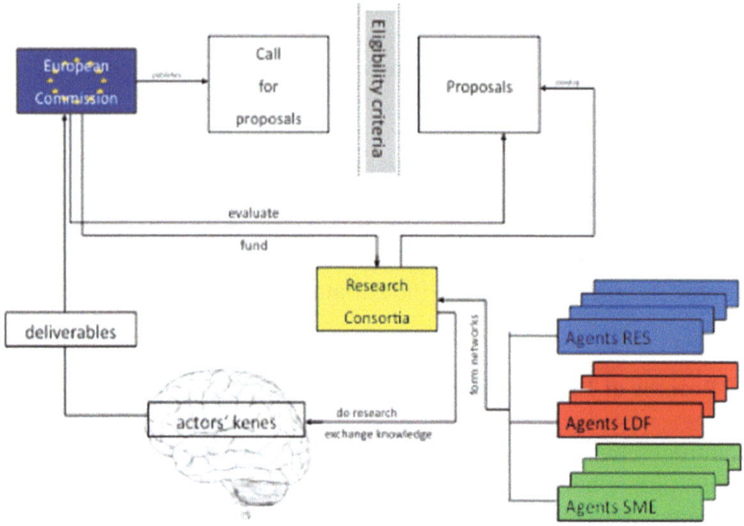

Fig. 7.1 Flowchart of INFSO-SKIN

The research undertaken in projects is incremental (abilities are substituted, expertise levels are increased). The potential of a radical innovation is determined only when the proposal is put together in the sense that new capability combinations can appear in consortia. SMEs are important candidates for contributing new capabilities and their involvement increases the likelihood for radical innovation. New knowledge is injected into the system most often by new, small and sophisticated companies. In order to allow the SMEs to play their special role we define ten capabilities per theme as "rare" capabilities and give these capabilities in the starting distribution exclusively to SMEs.

The learning processes and knowledge sharing in the projects follow the SKEIN version. The expertise levels of the capabilities used for the deliverables increase at each iteration. Capabilities of deliverables are exchanged among partners to model knowledge transfer in projects. At the end of the project all results are delivered to the Commission.

The flowchart above shows this modelling cycle of INFSO-SKIN (Fig. 7.1):

We define an iteration of the simulation to correspond to 1 month. This allows us to represent all processes in a reasonable order. For example, when a new Call is published, the deadline for a proposal is 6 months away, i.e. the agents have six time steps to set up a consortium and to "write a proposal". Another example is the length of research projects (e.g. 3 years, giving 36 iterations for consortium members to improve and exchange knowledge.)

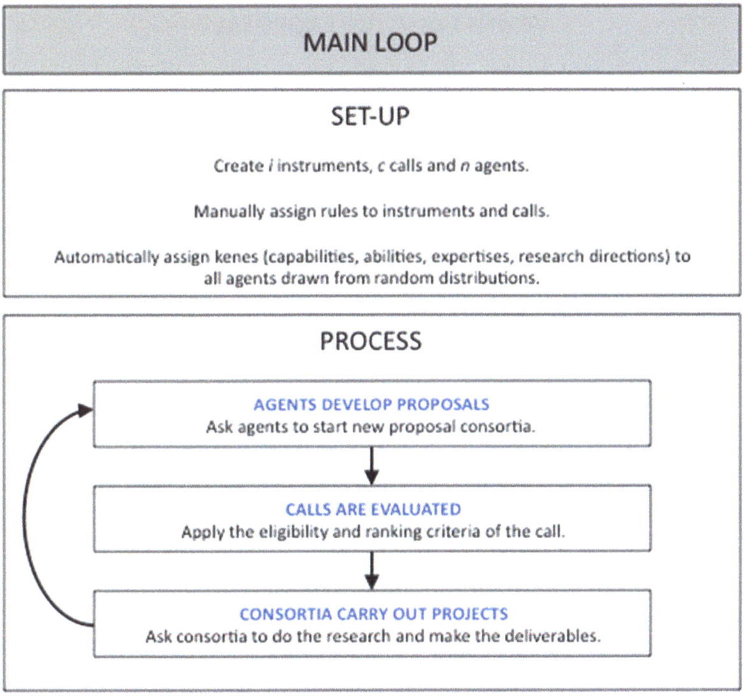

Fig. 7.2 Overview of the simulation model

7.2.2.5 Overview of the Main Processes

Figure 7.2 shows a flowchart of the above model narrative and Fig. 7.3 details each step.

7.2.2.6 Funding Decisions in the Model

The funding dimensions of the framework, instruments, calls, proposals and projects are implemented in the following way:

- For the Framework, a (global) value for total funding is set.
- For instruments, a multiplier is defined that relates to the average funding per partner per month. (The distribution of the index for FP7 is bell-shaped with equally long tails of "rich" and "poor" projects.)
- For Calls, the available funding is set as a percentage of the total Framework funding.
- For proposals and projects, the required EC contribution is calculated using the multiplier: $(contribution) = (multiplier) \times (size) \times (duration)$
- In the evaluation of proposals, the number of projects that will be funded depends on the sum of required contributions of the highest ranked eligible proposals and the available funding for the call.

AGENTS DEVELOP PROPOSALS

INITIATE NEW PROPOSAL

The possibility to initiate a proposal (most often by research institutes) depends on the agent's size (its capacity to develop new capabilities) and its existing projects.

FIND PARTNERS

The search process is guided by the requirements outlined in the call, which include a list of capabilities sought for proposal consortia.

- First, the agent looks on the list of his previous partners.
- Second, previous partners, who agreed to join the proposal, can add previous partners from their list.
- Third, new partners will be searched for.

These iterations are finite (a maximum search depth is set).

JOIN PROPOSAL

The possibility to join a proposal depends on the agent's size and its existing projects.

SUBMIT PROPOSAL

A proposal will be submitted when it fulfils the eligibility criteria. Otherwise the process is stopped and the agents of the consortium may start a new initiative.

CALLS ARE EVALUATED

APPLY ELIGIBILITY CRITERIA

Proposal consortia should fulfil all the eligibility criteria outlined in the call:

- The required composition of different types of agents;
- The desired research orientation, across the spectrum of basic and applied research;
- The minimum set of desired capabilities.

Consortia that do not fulfil these criteria are dissolved.

APPLY RANKING CRITERIA

All proposals that fulfil the eligibility criteria are ranked according to the rules:

1. The average expertise level of the proposal consortia;
2. The number of desired capabilities.

The Commission will support the *x* highest ranked proposals.
Consortia for lower ranked proposals are dissolved.

CONSORTIA CARRY OUT PROJECTS

INCREMENTAL RESEARCH

The research undertaken in projects is incremental research (abilities are substituted, expertise levels are increased).

MAKE DELIVERABLE

Consortium members are grouped in *s* sub-projects. Every *m* months the sub-projects produce a deliverable. This can be journal article or a patent.

LEARN FROM PARTNERS

Members obtain capabilities from partners in sub-projects. The capabilities that are learned are those contributed by the partners to the deliverable of the sub-project.

Fig. 7.3 Overview of the rules used in model's main processes

Table 7.2 Parameters for defining the Instruments

Min-number-of-partners	The minimum number of partners in project consortia
Max-number-of-partners	The maximum number of partners in project consortia
Composition	The types of partners in project consortia
Duration	The duration of projects in months

Table 7.3 Parameters for defining the Calls

Call-type	The type of Instrument that the Call is associated with (i.e. the type of Projects that the Call is for)
Call-publication-date	The publication date of the Call
Call-deadline	The deadline of the Call
Call-size	The number of Projects that will be funded
Call-orientation	The desired orientation of the research consortia, within the spectrum of basic research and applied research
Call-capabilities	A list of desired capabilities outlined in the Call
Call-capability-match	The required match between the capabilities contributed by project partners and the capabilities outlined in the Call

The Instruments are pre-defined at the model set-up. Differences between the Instruments are in the set of parameters that can be adjusted before running the simulation (Table 7.2).

The list of Calls is pre-defined at the model set-up, according to the list of Instruments. The set of parameters for each Call can be adjusted before running the simulation (Table 7.3).

In response to the (open) Calls, the agents create new proposals and develop their lists of partners and previous partners. Meanwhile (and as result of their interactions), the agents develop their kenes. The model's interface provides information about the overall collaboration network of the programme, the nodes being the organisations that are beneficiaries of funding and the edges are the links to their partners. This network has certain structural properties (Table 7.4).

In addition to these measures about the collaboration network, the model's interface provides several statistics on the participants, proposals and projects over the course of the simulation (Table 7.5).

The statistics on project consortia provided by the model's interface (Table 7.6) help to interpret the collaboration network measures (e.g. a high percentage of small projects can explain a low network density).

To find a realistic starting configuration for simulating EU-funded ICT R&D networks, we calibrated the model using an existing database of ICT project networks funded by DG INFSO. Figure 7.4 shows the information from the DG INFSO empirical database that we used for the model.

Table 7.4 Measures calculated on the collaboration network

Number-of-nodes	The number of nodes (agents)
Number-of-edges	The number of edges (representing collaborations between agents)
Network-density	The proportion of edges in the network relative to the total number possible (sparse versus dense networks)
Number-of-components	The number of components (i.e. portions of the network in which all agents are connected, directly or indirectly, by at least one link)
Size-of-largest-component	The size of the largest ('giant') component
Average-degree	The average number of connections a node has to other nodes
Diameter	The maximal distance between any two nodes in the network
Average-path-length	The average number of steps along the shortest paths for all possible pairs of network nodes
Clustering-coefficient	The degree to which nodes in the network tend to cluster together ('cliquishness')

Table 7.5 Statistics on participants

Participation-in-proposals-distribution	Frequency distribution of the number of Proposals that agents are currently in
Participation-in-projects-distribution	Frequency distribution of the number of Projects that agents are currently in
Degree-distribution	Frequency distribution of the number of partners (or previous partners) that agents currently have in their list

Table 7.6 Statistics on project consortia

Projects-count	The number of projects supported by the Commission, totalled over all the 6 Calls
Projects-with-SME-%	The percentage of projects with 1 or more SME partner
Projects-size-statistics	The average, median, min and max size of project consortia, as well as the % of 'small' project consortia (less than 5 partners) and the % of 'big' consortia (more than 20 partners)

7.2.2.7 Creating the Starting Configuration

The starting configuration of the model is constructed using the DG INFSO database. First, a list of projects from the first FP7 call is retrieved from the database (IP or STREP). Agents (research institutes, diversified firms and SMEs) are then created, based on the organisations that are listed as participants of these projects. Links are created between all the members of the projects. More agents are then created up to a user-specified number, for example 3,000 agents, which will be larger than the number of organisations in the FP7 database. This is the case to give the opportunity that the network can grow by connecting so-far not connected agents to the existing (starting) network. It mirrors reality where there are many non-EU-funded research organisations out there which, however, nevertheless

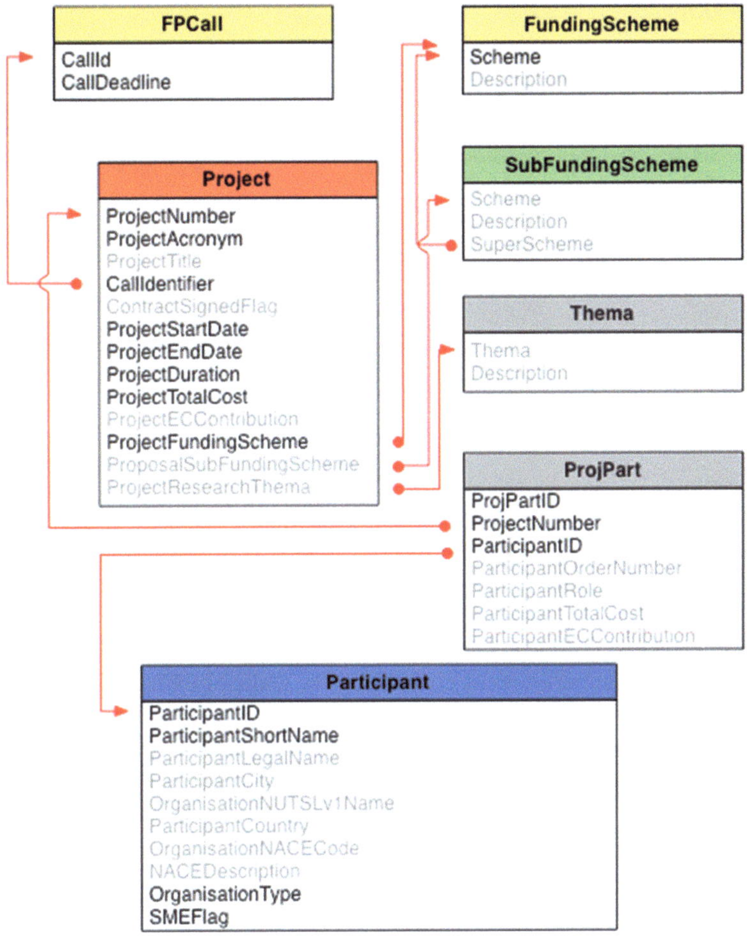

Fig. 7.4 Database information from DG INFSO for starting networks in INFSO-SKIN. *Highlighted items* were used to calibrate the simulation

write proposals and—if successful—newly connect to the network in the consequent calls. We need to address the issue that some participants stay under the radar, despite their attempts to enter the network by writing proposals. Since we did not get any data from DG INFSO about the ones trying but not succeeding, we can only estimate a number (3,000) that is big enough to allow the network to grow. These additional agents are not part of the starting network.

Participant size, a measure of the capacity of organisations to interact and exchange knowledge, is based on total FP7 funding, information read from the DG INFSO database. If participants are very big, multiple agents (proportional to the funding) will be created to represent the organisation. Agents are endowed with their initial kenes (capabilities, abilities, expertise levels and research directions) using randomised designs for the agent types, based on the model's partitioning

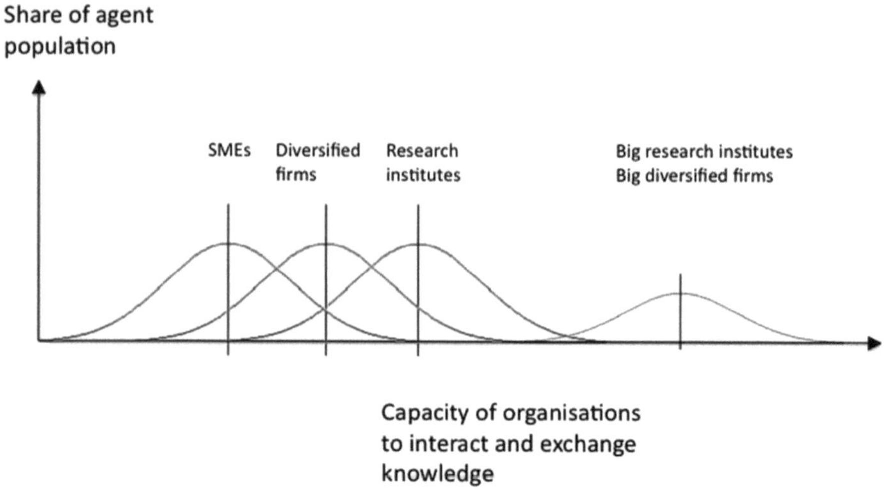

Fig. 7.5 Configuration of initial capacity of organisations

(with nine themes, common/rare capabilities) of the knowledge space, in a way that results in a distribution similar to that shown in Fig. 7.5. We need a partition of nine to carry out the first policy change experiment, i.e. extending the thematic areas by one. Eight is the number of areas covered by the Calls of FP7, the ninth would not be covered by the Calls but addressed under the first experiment for Horizon 2020. So it needs to be there "empty" (=un-addressed) as a possibility from the beginning. The ninth thematic area is unaddressed in our experiments, except in the one experiment that explores the impact of extending the number of areas to 9.

Instruments and Calls are then created to complete the setup procedure.

7.2.3 The Baseline Scenario as Benchmark

For the Calls 1–6 that has actually occurred in FP7, we have data on the number of participants (funded organisations differentiated between research institutes (RES), large diversified firms (LDFs) and small and medium sized enterprises (SMEs)), and the number of funded projects. Furthermore, we have information on the size of the projects as measured by the number of participants, the duration of the project, and the average funding per project. Analysing this information provides empirical information on important network indicators. These numbers can be used to calibrate and validate the model.

Table 7.7 Results of baseline scenario per instrument (IP)

			Sim IP (15 runs)	
		Emp IP	Mean	St dev
Participants	RES in projects	599	569.7	16.572
	LDFs in projects	458	501.3	14.105
	SMEs in projects	462	520.1	18.928
	Participants in proposals (avg)	**N/A**	**2.578**	**0.083**
	Participants in projects (avg)	1.968	2.002	0.030
Proposals	**Number of proposals**	**N/A**	**347.5**	**19.276**
	Proposals-size-avg	**N/A**	**13.682**	**0.187**
	Proposals-capability-match-avg	**N/A**	**19.4**	**0.229**
Projects	Number of projects	209	206.9	7.652
	Consortium size (avg)	15.383	13.622	0.213
	Project duration (avg)	41.2	41.0	0.000
	Project funding (avg)	7.9	7.6	0.118
Knowledge	**Knowledge per participant**	**N/A**	**16.4**	**0.054**
	Knowledge flow per project	**N/A**	**13.8**	**0.221**
Capabilities	**Capability diffusion (Theme 1–8)**	**N/A**	**0.729**	**0.006**
	Capability diffusion (Theme 9)	**N/A**	**0.706**	**0.007**
	Capability frequency (avg)	**N/A**	**54.6**	**0.178**
Network	Density	0.019	0.018	0.001
	Number of components	1	1.0	0.000
	Size of the largest component	1,634.0	1,634.2	37.474
	Diameter	4	4.067	0.258
	Average-path-length	2.539	2.620	0.017
	Average-degree	30.7	30.2	0.532
	Average-clustering	0.840	0.728	0.007

N/A not available (unobservable)

7.2.3.1 FP7 Simulation

Comparing the empirical and—for the Baseline Scenario—the simulated data, the next two tables below summarise all values per instrument (IP and STREP). The values where we have empirical information are given in black. Values for the empirical "unobservables" (proposal writing activities, the amount of knowledge held by the agents, and knowledge flows between the agents in the network) are in bold and obtained only from the simulation. Because several stochastic components are included in the model, we chose a modified Monte-Carlo approach and repeated the experiments 15 times. The choice of 15 runs for simulation of the baseline scenario is founded on visual inspection of the plots of convergence for the indicators in Tables 7.7 and 7.8, which show that 15 replications are enough for a reliable and stable result. In the result tables we display average values. Scatter plots also show that this number of repetitions produces reasonably smooth results and excludes outliers due to stochastic fluctuations.

Table 7.8 Results of baseline scenario per instrument (STREP)

		Emp STREP	Sim STREP (15 runs) Mean	St dev
Participants	RES in projects	786	809.9	12.738
	LDFs in projects	522	692.3	8.464
	SMEs in projects	747	804.6	11.005
	Participants in proposals (avg)	**N/A**	**4.022**	**0.095**
	Participants in projects (avg)	2.354	2.340	0.021
Proposals	**Number of proposals**	**N/A**	**1,292.1**	**35.789**
	Proposals-size-avg	**N/A**	**8.231**	**0.056**
	Proposals-capability-match-avg	**N/A**	**11.0**	**0.068**
Projects	Number of projects	644.0	592.3	5.496
	Consortium size (avg)	8.025	8.208	0.072
	Project duration (avg)	34.2	34.0	0.000
	Project funding (avg)	2.8	3.0	0.028
Knowledge	**Knowledge per participant**	**N/A**	**17.1**	**0.048**
	Knowledge flow per project	**N/A**	**6.8**	**0.145**
Capabilities	**Capability diffusion (Theme 1–8)**	**N/A**	**0.745**	**0.010**
	Capability diffusion (Theme 9)	**N/A**	**0.704**	**0.006**
	Capability frequency (avg)	**N/A**	**57.0**	**0.160**
Network	Density	0.008	0.009	0.000
	Number of components	2	1.0	0.000
	Size of the largest component	2,190.0	2,349.7	18.495
	Diameter	6	5.000	0.000
	Average-path-length	2.846	2.846	0.015
	Average-degree	17.0	20.4	0.176
	Average-clustering	0.812	0.603	0.005

N/A not available (unobservable)

7.2.3.2 Calibration

Tables 7.7 and 7.8 confirm that the model achieves a good match between the empirical and the simulated number of participants and projects. We defined a corridor for our simulation parameters where the simulated numbers sufficiently match the empirical numbers. This gave us the following distributions of number of participants per project (project size) and funding per instrument (project contribution) for our Baseline Scenario (Fig. 7.6) including the empirical distributions.

Note the empirical presence of very large IP projects in the first of these four figures. We considered them to be a consequence of administrative actions by the Commission, merging certain projects, not an endogenous outcome.

The figures illustrate the distributions of the number of actors in consortia (project size) and the financial size of projects (project contribution). In all cases we also observe a few larger consortia as well as financially larger projects. Using

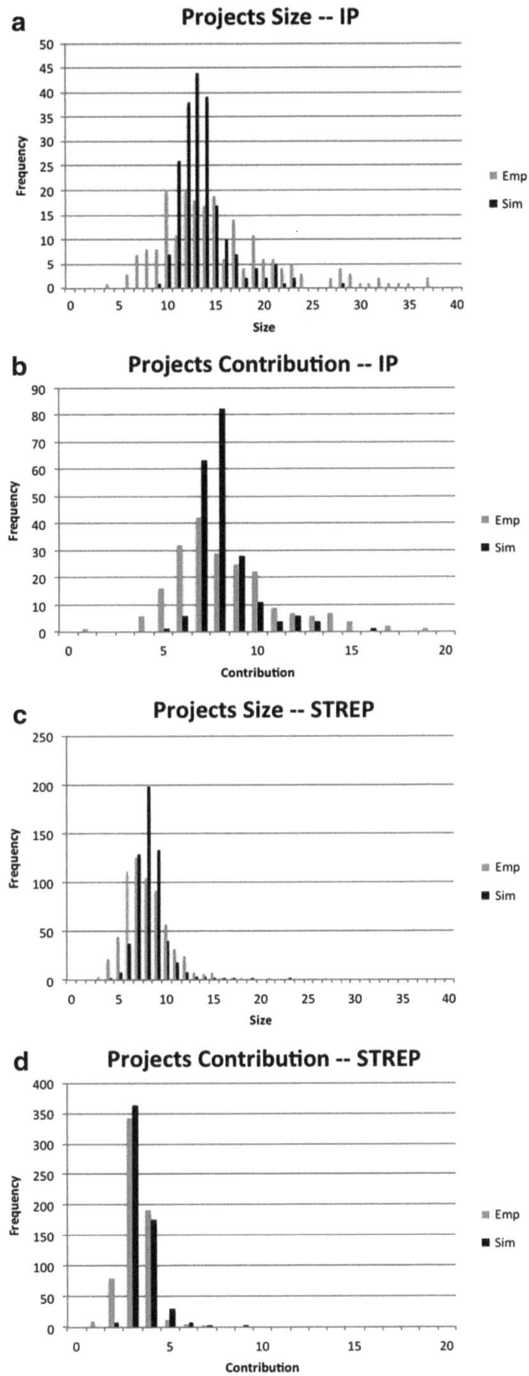

Fig. 7.6 Distributions of projects size and funding per instrument (IP and STREP) in the baseline scenario

the empirical information on participants and projects matched by the calibrated simulation, the basic network indicators could be reproduced. The ones in the tables above are explained in more detail here.

Why does the empirical information looks similar to the simulated one for the starting period, which expresses a delay of responding, which would only be valid for the simulation data? The short answer is that this is purely a consequence of the way of reporting the model results, as explained below. Because the empirical data is reported in exactly the same way (by running the model in the empirical case mode) the similarity in curves is to be expected.

To go into more detail: The internal representation of agents and the external reporting of the agent world are separate matters. Internally, each agent has two lists: one list with its current partners and one list with its previous partners. When a project is completed, the partners are moved from the current partners list to previous partners list, iterating over each member of the consortium. Externally, we had to make a choice how to report the network evolution, since it is possible to concentrate on the network of current partners or on the network of previous partners. This is merely a reporting choice that does not have any consequences on the model results themselves: internally, the two above-mentioned lists are the only information available and used about the network among participants.

Moreover, the reporting choice we made during the study is to concentrate on the previous partners network because we want to show the growth of the network over time. This choice has indeed introduced a reporting delay and we only see growth of the network after completion of some of the initial projects. The size of the delay is to be gotten from Tables 7.7 and 7.8, which contain the average duration of IP (41.2 months) and STREP (34.2 months) projects. These numbers, plus that many of the first completed projects have started at some time $t > 1$, account for the delay visible.

In hindsight, the whole delay issue could have been avoided by reporting the current **and** previous partners network, i.e. the sum of the network of current partners and the network of previous partners. However, our reporting choice (and the consequent delay) is immaterial for calibration and evaluation, since it has been consistently applied to both the empirical data and the simulation data, plus the fact that we used only the end values ($t = 112$) for doing the t-tests, which makes it even less of an issue.

The development of the empirical largest component and of the largest component of the baseline scenario over the 122 months of the simulated Programme are displayed in Fig. 7.7. This indicator comprises the number of agents who can in principle be reached by traversing links between agents within the largest coherent part of the network. One of the basic aims of Framework Programmes is to contribute to the establishment of a European research landscape and the size of the largest component can be interpreted as a measure of the achievement of this goal. As can be seen from the figure, the largest component remains at the initial size for $36 + 6$ months and then starts to increase because of the delay in reporting: participants announce their updated partner lists only once a

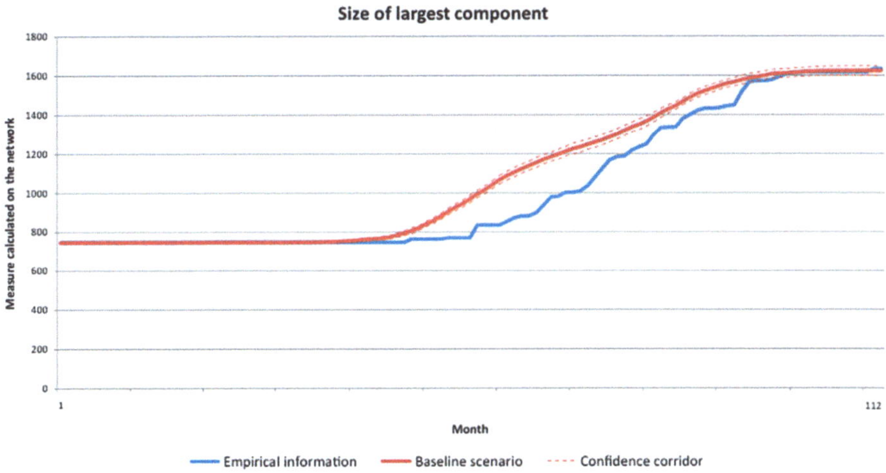

Fig. 7.7 Development of the largest component

project has finished, which correspond to 6 months proposal writing period plus 3 years project work after the first Call has been published.

The second indicator is the network density, which describes the overall level of linkage among the actors. It is the number of actors who are connected to each other, expressed as a percentage of the maximum possible number of connected actors. The indicator measures the effects of policy interventions on the extent and speed of knowledge diffusion in the network. Higher density values indicate faster potential diffusion of knowledge; lower values indicate a reduction in the speed of diffusion. Although the interpretation of the density is straightforward, one has to keep in mind that it is a relative measure and its magnitude depends on the overall network size. Therefore, densities of networks of different sizes cannot be compared. As indicated by the growth of the largest component we are confronted with growing networks, which lead to a decreasing density value over time (Fig. 7.8).

The average degree and the degree centrality are widely used indicators from SNA, especially in cases where large networks of varying sizes are analysed. The average degree simply divides the sum of the number of connections of every agent by the number of agents and gives a rough idea of the overall direct connectivity of the agents, an indicator of the cohesion of the network. Figure 7.9 shows the empirical and simulated development of the average degree. Because the degree is increasing in later periods we conclude that the cohesion of the network increases despite the decreasing density.

However, the stage of development of the network's knowledge base also matters. In early explorative stages the average degree might decrease because new actors bring in new knowledge fields; for more mature knowledge bases, a decreasing average degree might indicate a dissolving network. The structure of the research networks referring to the various knowledge fields and their connections can be approximated by the average clustering coefficient of the network. Higher

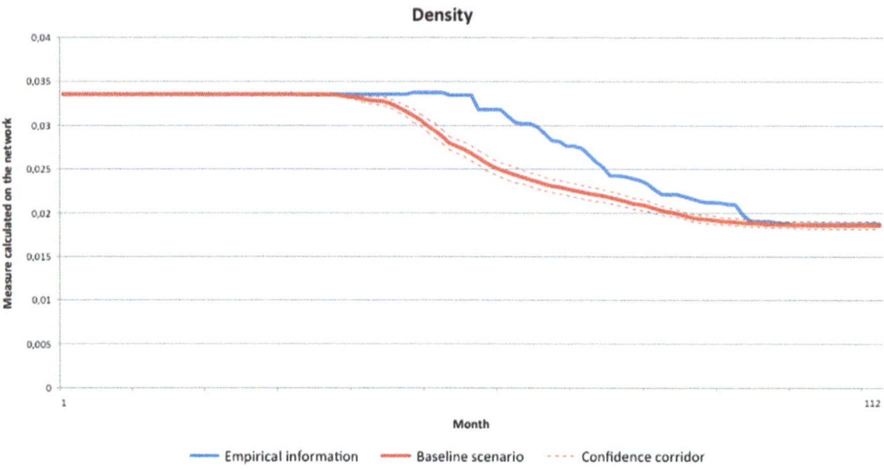

Fig. 7.8 Development of the network density

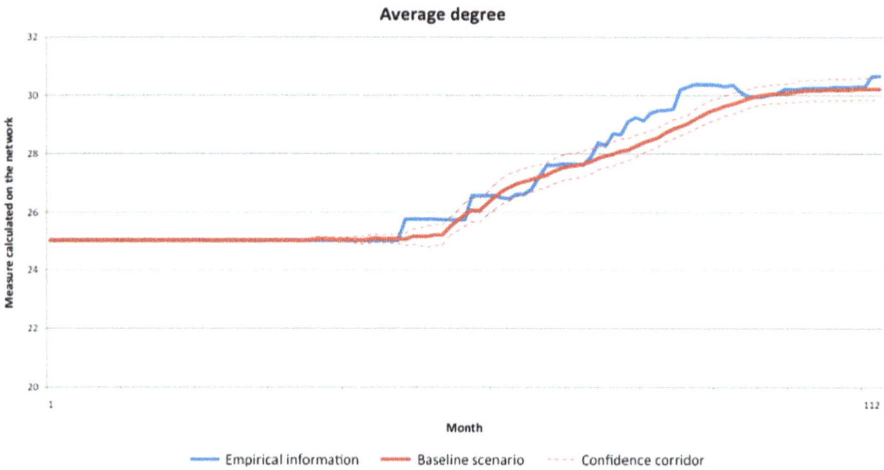

Fig. 7.9 Development of the average degree

clustering coefficients indicate that research networks are characterized by highly connected subgroups of agents that are only loosely connected by few linkages. For such networks, critical actors who might act as boundary spanners can be critical for the transmission of ideas. Figure 7.10 illustrates that in the simulated networks, clustering is decreasing and therefore structural holes between different knowledge areas are increasingly closed.

The average path length is defined as the average number of steps along the shortest paths between all possible pairs of agents in a network. In the case of knowledge transfer, the quality of knowledge exchange processes is negatively

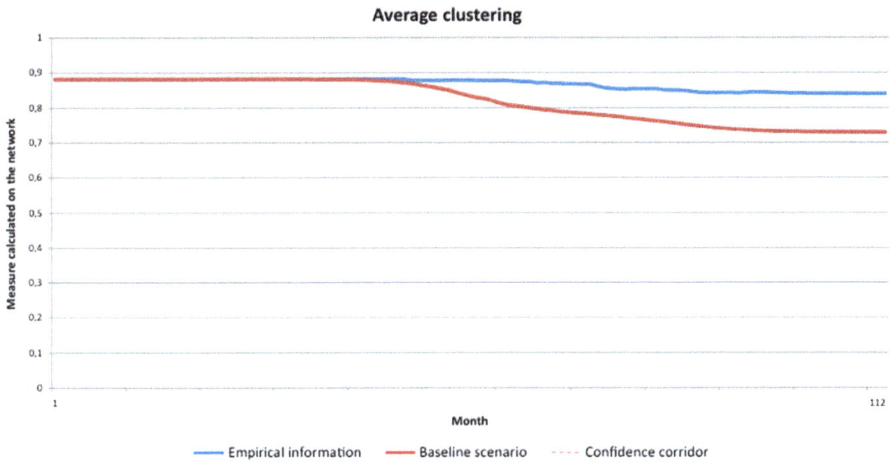

Fig. 7.10 Clustering in the networks

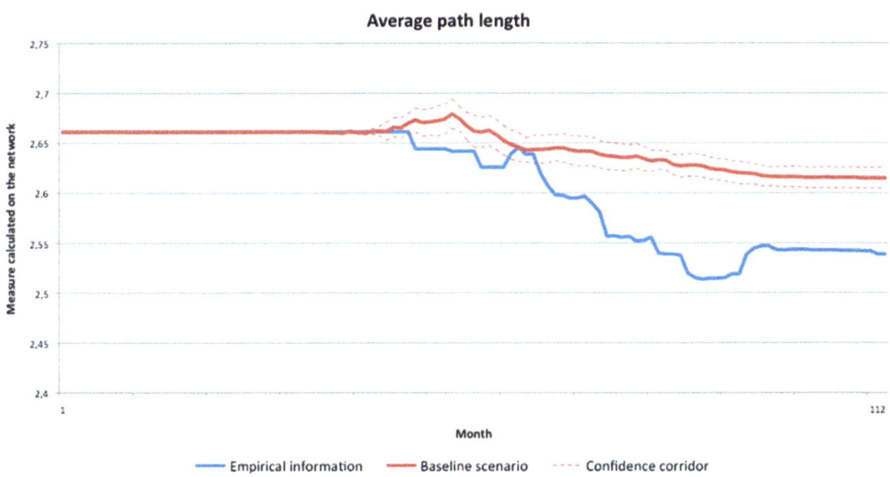

Fig. 7.11 Average path length

affected by increasing path length. Figure 7.11 displays for the baseline scenario as well as for the real network the development of the average path length. In both cases, we find a decrease in average path lengths over time. This result is to be interpreted as a success for network policies in FP7. In Fig. 7.11, the empirical trend is comparatively stronger; the difference between this and the simulation results can be traced back to the construction of the database, which only includes actors in consortia, whereas the simulated data also includes unconnected actors, leading to the relatively higher values of this indicator.

As one of the "un-observables", we record for the Baseline Scenario the number of proposals, how many participants are involved in proposal writing activities, and how many are actually engaged in projects. In the tables for the experiments, there are two indicators listed: the actual number of proposals (*proposals*), and the thematic match between proposal requirements in terms of knowledge capabilities and the capabilities recorded in the submitted proposals (*cap-match*).

The number of proposals produced in the simulation is double the number of projects that are funded. This means a model success rate for proposals (IP and STREP) of approximately 50 %, which is well above the empirical success rate of 20 % according to FP7 subscription statistics (source: *FP7 Subscription, Performance, Implementation report, 2007–2008*). However, in this case we do not need an empirical match, because the agent population in the simulation does not mirror the number of all potentially participating organisations in the European funding landscape that could theoretically submit proposals. We just need a realistic evaluation procedure for submitted proposals to turn them into accepted proposals (=funded projects) where the number of accepted proposals, i.e. projects, matches the project number empirically recorded. The acceptance conditions for proposals have been set to match as closely as possible the empirical size distribution of projects.

The knowledge-related indicators of the Baseline Scenario (Ahrweiler et al. 2011) demonstrate the additional advantage of a simulation to record the development of another "un-observable". Empirically, we cannot see and only with difficulty can we measure knowledge generation, knowledge exchange, and learning. In the simulation, the effects of knowledge exchange processes are measured as knowledge flows between the agents in the networks. To measure the amount of knowledge of participants, we used a measure (k) that is equal to the length of a participant's kene. The average knowledge per participant is:

$$\overline{k} = \frac{1}{nParticipants} \sum_{i \in \{Participants\}} k_i$$

and the standard deviation of knowledge among the participants is

$$\sigma_k = \sqrt{\frac{1}{nParticipants} \sum_{i \in \{Participants\}} \left(k_i - \overline{k}\right)^2}$$

The knowledge flow is the total increase of knowledge over all participants arising from their learning from partners. Let k_i^{t-1} be the knowledge of a participant before learning from partners and k_i^t be the knowledge of a participant after learning from partners. The knowledge flow then is

Table 7.9 Information needed for computing the diffusion of capabilities

Capabilities	The overall number of capabilities
Capabilities-diffusion	The share of participants carrying a capability from a certain theme (see Fig. 7.12)
Capabilities-frequency	The presence of capabilities: how many participants of a certain type have a certain capability?

$$kf = \sum_{\{\text{Projects}\}} \sum_{i \in \{\text{Partners}\}} \left(k_i^t - k_i^{t-1}\right)$$

An additional indicator focuses on the qualitative composition of the networks' knowledge bases. Modifying the call for proposals and asking for different knowledge areas to be involved might exert an influence on the depth and breadth of the agents' knowledge base. This is measured with the number of shared capabilities among the agents. We implemented an indicator (*diffusion*) to track the diffusion of capabilities based on the information in Table 7.9.

$$D_x = \frac{|\{(i) : i \in \{Participants\}, X_i \cap X_k \neq \emptyset\}|}{|\{Participants\}|}$$

where

D_k is the diffusion value for capabilities from theme k

X_i is the set of capabilities carried by participant i

X_k is the set of capabilities from theme k

Figure 7.12 shows the Baseline Scenario values for the diffusion of capabilities'. Note the lower diffusion value for capabilities from theme 9, which are not outlined in the Calls.

7.2.3.3 Replications and Confidence Limits

The number of replications used for calculating statistics about the Baseline Scenario and each of the experiments is 15. This is sufficient for convergence and yields a small enough confidence interval for the set of indicators used (see Fig. 7.13). The issue of the number of simulation replications is covered in (Chap. 9).

The simulated data for the Baseline Scenario is summarized using corridors around the mean (Fig. 7.14). The plots for each of the 15 replications for most part will lie within this corridor. Due to the small number of replications, the corridors were constructed using t-based confidence intervals (95 %). Note that the summarizing tables in this study only present the end value (after 112 months) and the standard deviation.

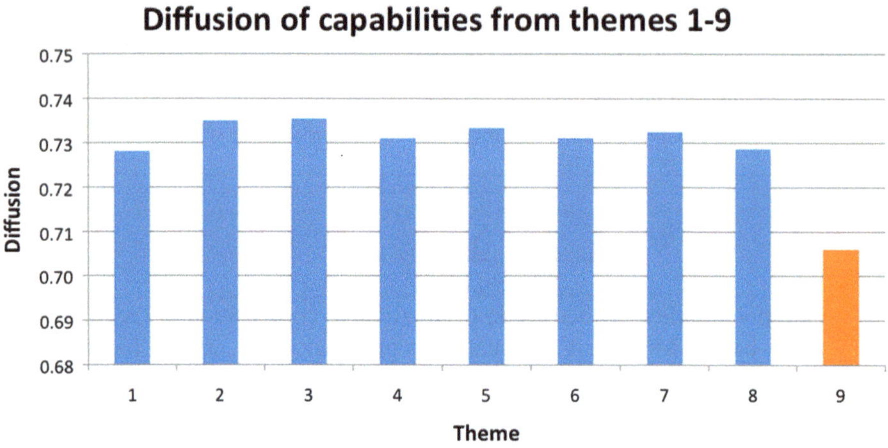

Fig. 7.12 Diffusion of capabilities in the baseline scenario

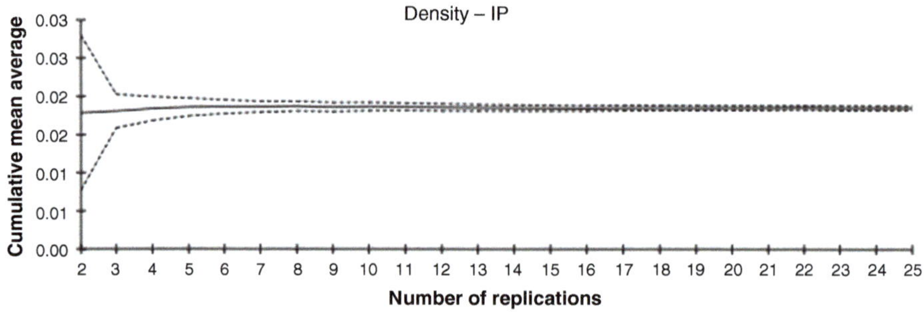

Fig. 7.13 An example of convergence and confidence intervals (shown for the density indicator, baseline scenario, IP)

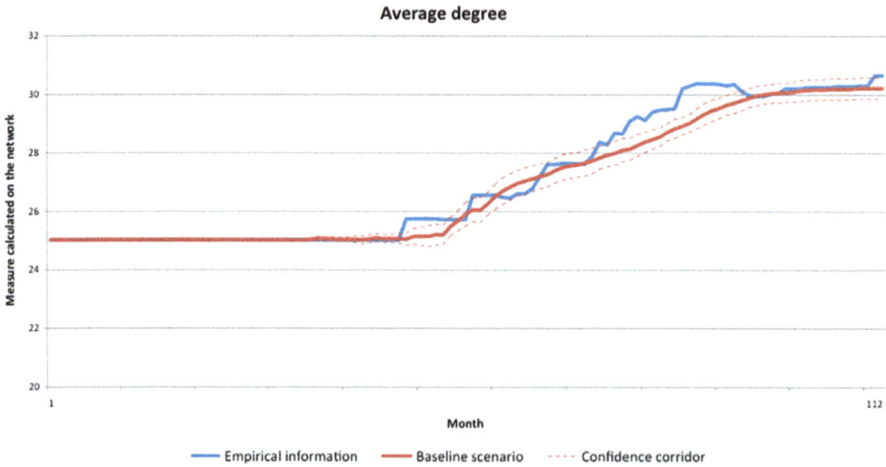

Fig. 7.14 Corridor for simulated data (shown for average degree indicator, baseline scenario, IP)

Table 7.10 Results of the baseline scenario per instrument (IP)—Horizon 2020

		Sim IP (15 runs)		Sim IP 2020 (15 runs)	
		Mean	St dev	Mean	St dev
Participants	RES in projects	569.7	16.572	625.0	17.790
	LDFs in projects	501.3	14.105	568.6	13.323
	SMEs in projects	520.1	18.928	608.8	19.693
	Participants in proposals (avg)	2.578	0.083	3.836	0.110
	Participants in projects (avg)	2.002	0.030	2.521	0.060
Proposals	Number of Proposals	347.5	19.276	621.9	32.338
	Proposals-size-avg	13.682	0.187	13.896	0.134
	Proposals-expertise-level-avg	5.1	0.053	4.8	0.053
	Proposals-capability-match-avg	19.4	0.229	19.4	0.116
Projects	Number of projects	206.9	7.652	311.4	7.511
	Project size (avg)	13.622	0.213	13.739	0.159
	Project duration (avg)	41.0	0.000	41.0	0.000
	Project funding (avg)	7.6	0.118	7.6	0.088
Knowledge	Knowledge per participant	16.4	0.054	14.5	0.055
	Knowledge-flow per project	13.8	0.221	15.5	0.450
Capabilities	Capability diffusion (Theme 1–8)	0.729	0.006	0.684	0.010
	Capability diffusion (Theme 9)	0.706	0.007	0.634	0.008
	Capabilities frequency (avg)	54.6	0.178	48.3	0.182
Network	Density	0.018	0.001	0.019	0.001
	Number of components	1.0	0.000	1.0	0.000
	Size of the largest component	1,634.2	37.474	1,845.4	40.540
	Diameter	4.067	0.258	4.040	0.200
	Avg-path-length	2.620	0.017	2.546	0.017
	Avg-degree	30.2	0.532	35.0	0.669
	Avg-clustering	0.728	0.007	0.668	0.008

7.2.3.4 The "Business as Usual" Experiment

The discussion above show that the networks created by DG INFSO in FP7 and those created by the agent-based model are qualitatively similar. The simulation experiments can be characterized as *history-friendly* experiments, which reproduce the empirical data and cover the decisive mechanisms and resulting dynamics of the real networks. This qualifies the model to be applied for *in silico* experiments, which take these outcomes as a starting point and test potential future developments in the sense of the *ex-ante evaluation* outlined above. The first experiment simply prolongs the time horizon and keeps the instrument design as outlined in FP7. In other words the coming funding programme, Horizon 2020, is treated as a simple extension of FP7.

The outcomes of this experiment, which we have labelled "Business as Usual", are outlined in Tables 7.10 and 7.11, one for the IP and one for the STREP case. We are interested in the effect on the number and composition of participants, the number of submitted proposals and the qualitative features of the underlying network structures. Analysing these figures allows for an ex-ante evaluation of FP7 instruments concerning their sustainability beyond the FP7 time horizon.

Table 7.11 Results of the baseline scenario per instrument (STREP)—Horizon 2020

		Sim STREP (15 runs)		Sim STREP 2020 (15 runs)	
		Mean	St dev	Mean	St dev
Participants	RES in projects	809.9	12.738	855.6	13.166
	LDFs in projects	692.3	8.464	734.3	8.834
	SMEs in projects	804.6	11.005	878.5	14.633
	Participants in proposals (avg)	4.022	0.095	6.560	0.139
	Participants in projects (avg)	2.340	0.021	2.907	0.031
Proposals	Number of proposals	1,292.1	35.789	2,286.3	51.279
	Proposals-size-avg	8.231	0.056	8.304	0.032
	Proposals-expertise-level-avg	5.1	0.028	4.8	0.034
	Proposals-capability-match-avg	11.0	0.068	11.1	0.049
Projects	Number of projects	592.3	5.496	822.9	5.456
	Projects-size-avg	8.208	0.072	8.199	0.054
	Projects-duration-avg	34.0	0.000	34.0	0.000
	Projects-contribution-avg	3.0	0.028	3.0	0.020
Knowledge	Knowledge per participant	17.1	0.048	15.2	0.050
	Knowledge-flow per project	6.8	0.145	7.8	0.151
Capabilities	Capability diffusion (Theme 1–8)	0.745	0.010	0.704	0.009
	Capability diffusion (Theme 9)	0.704	0.006	0.633	0.007
	Capability frequency (avg)	57.0	0.160	50.8	0.167
Network	Density	0.009	0.000	0.009	0.000
	Number of components	1.0	0.000	1.0	0.000
	Size of the largest component	2,349.7	18.495	2,511.4	23.353
	Diameter	5.000	0.000	5.000	0.000
	Avg-path-length	2.846	0.015	2.768	0.013
	Avg-degree	20.4	0.176	23.4	0.273
	Avg-clustering	0.603	0.005	0.554	0.006

At first glance, the results of the *Business as Usual Experiment* confirm the sustainability of FP7 policy designs. Extending policies beyond FP7 can be considered as a strategy that conserves the trends sustaining the networks. A closer look at the relevant figures, nevertheless, indicates some subtle changes that qualitatively affect the networks and the resulting dynamics.

All the agent groups (RES, LDF and SME) are affected, i.e. their rate of participation grows, though the rates differ. The lowest growth rate of approximately 10 % is for research institutes, perhaps because the majority of institutes are already in the networks. Large Diversified Firms (LDFs) are the second fastest growing group and extend their participation by approximately 13 %. Again, most of the European LDFs, at least those including R&D departments, are already linked to the networks. The highest growth in participation is for Small and Medium Sized Enterprises (SMEs), which increase their participation on average by 17 %.

The strongest effect is on the number of proposals being submitted, which almost doubles in the extension experiment. There is no change in the average number of agents that join a proposal team. However, the average expertise characterising proposal writers decreases moderately, as a side-effect of the increasing proposal activity: agents with somewhat lower expertise levels manage to be included in proposal teams because of the scarcity of relevant competences. This can be considered as a factor supporting SMEs relative to LDFs and RESs. The number of proposals, which succeed in passing the evaluation of the Commission also increases, but only by half as much as the growth in the number of proposals. The increase in the rejection rate implies an increase in the quality of the projects. There are no significant changes to the average project size, duration or Commission contribution (the value of financial support per project).

The quantity and quality of linkages among project consortium members moderately increases, which leads to increasing knowledge flows within the projects. With the larger number of agents in the networks, a decrease in the rate of knowledge diffusion was to be expected and, indeed, the capability diffusion slightly decreases. In total, however, the conditions for knowledge flows within the network and therefore for mutual learning and cross-fertilisations increase: On the one hand, the largest component—measuring the number of agents that are in principal connected in the network—increases by 12 %, complemented by a moderate increase in the network density. The increase in the average degree from 30 to 35 reflects a more egalitarian distribution of linkages caused by a higher participation of SMEs and a higher number of projects. These developments lead to a decrease in the clustering in the network, which corresponds to a somewhat decreasing influence of LDFs.

To summarise, *Business as Usual* showed that under a continuation of FP7 policies networks not only survive but also grow, while the structure of those networks tends to evolve towards a more balanced distribution of influence between agents, and the networks become more effective at producing proposals.

7.3 Experiments with Policy Changes

In helping to design the new Horizon 2020, DG INFSO had the opportunity to consider new policy options such as switching to a stronger support for certain funding instruments, to accentuate agenda-driven aspects, or to address new topical areas in order to initiate the emergence of new industries and scientific fields.

The following policy changes were tested for ex-ante evaluation of the new programme: (1) What if there would be changes to the thematic scope of the programme? (2) What if there would be changes to the instruments of funding? (3) What if there would be changes to the overall amount of programme funding? (4) What if there would be changes to increase SME participation?

7.3.1 The Organisation of the Experiments

To illustrate results of the different experiments and demonstrate the effects on the various targets, we can only provide a very short introduction here and refer the reader to a different publication (Ahrweiler et al. 2014a).

The results of the experiments have to be understood against the baseline scenario, the reproduction and dynamic extension of FP7 (Sect. 7.2.3.4 above). The comparison of outputs (baseline scenario and the scenario under investigation) was checked with an independent two-sample t-test with equal sample sizes and unequal variance. Comparisons that are statistically significant are those with values less than 0.05.

There are many dimensions of the network evolution that can be used for the comparisons. In order to cover the multi-facetted aspects of our *explanandum*—the networks and their dynamics—we tabulate the four experiments and the effects of the modified policy design in six rows (see Table 7.12). The heads of the columns label the different experiments performed, and the rows are labelled with the variables under investigation.

In the first row the effect on the number of participants and their rate of participation in the networks is listed. In particular we differentiate between research institutes (RES), large diversified firms (LDFs) and small and medium sized enterprises (SMEs). An increasing rate of participation is indicated with a plus, while a minus indicates decreasing participation rates.

The second row lists the number of proposals, and whether there is an increase in proposal writing activities (plus) or a decrease (minus). This figure is of interest as the number of proposals submitted can be interpreted as an indicator determining the potential for network growth.

In the third row, the effects on the projects are listed. Especially, we differentiate between the number of projects, the size of the projects as measured by the number of participants, the proportion of each agent type, and the average funding per project. The respective figures can be used to evaluate the effects of various sizes and forms of project consortia on the overall network evolution.

The fourth row deals with the effects of knowledge exchange processes measured as knowledge flows between the agents in the networks. The modifications in the various experiments can lead either to increasing or decreasing mutual learning triggered by knowledge flows and therefore allows for the evaluation of the effectiveness of various policy designs for knowledge diffusion.

The last row deals with the network as a whole. To illustrate deviations from the baseline scenario, the effects of the experiments on important network indicators are documented in the table and their development over time is displayed.

Any statistically significant deviation from the Baseline Scenario for these indicators is marked by a plus (+) or a (−). No significant effect is recorded by the entry “no effect” in the respective cell.

Table 7.12 Possible ranges of the results of experiments

	Programme scope experiment		Hubs and spoke experiment				Elite and selectivity experiment		Specialisation experiment	
	A Lower number of thematic areas	B Higher number of thematic areas	C Smaller consortia	D Larger consortia	E Higher funds per consortium	F Lower funds per consortium	G Higher funds per program	H Lower funds per program	I More SMEs	J More applied research
Participants	SMEs (−0+) LDF (−+0) RES (−0+)									
Proposals	More submitted proposals (+) Less submitted proposals (−)									
Projects	Size of consortia (+0−); participation of SMEs, RES, LDF (+0−)									
Knowledge	Increasing (+) or decreasing (−) knowledge exchange among agents									
Capabilities	Wide (+) or narrow (−) diffusion of capabilities									
Networks	Density (−0+), Average Path Length (−0+), Average Degree (−0+), Average Centrality (−0+), Average Clustering (−0+)									

7.3.2 Summary Results

The "Programme Scope" experiment indicated that new thematic areas could be easily integrated (or existing ones removed) without very much effect on most parameters. These changes would not impact significantly on the overall network structure, which is described by the number of participants and the number of projects. The thematic scope of the programme would basically be a political decision. It is a decision on the direction and desirable focus of research. Given that the intended changes remain within a non-critical corridor, it is not likely that the inclusion or exclusion of themes will cause problems for the achievement of targets such as density and size of the networks.

The "Hub and Spokes" experiment shows the positive effects on the research networks arising from smaller consortia sizes and to some extent also lower rates of monthly funding. Huge projects with very high funding do not support the network-related goals. Of course, other aims in the underlying research policy might require policy designs favouring larger projects.

In the "Elite or Selectivity" experiment, decreasing funding leads to a higher degree of competition for the potentially smaller number of projects. A larger number of proposals are submitted, but only a smaller number of consortia will be successful. This can be interpreted from the evaluator's side as being more selective and demanding a higher quality of project proposals.

In the "Specialisation" experiment, SMEs were very successful with their highly specialised technological knowledge bases; they are well represented within the projects. This suggests that having more SMEs in the overall agent population would in itself be a good trigger for increasing the participation of SMEs because they tend to be successful in the competitive process of proposal evaluation. Also, policy changes to a more applied research direction would be helpful for more SME participation.

7.4 Conclusions

To what degree can policy decisions be made on the basis of these conclusions, i.e. how accurate is the model? The model was calibrated from the Calls 1–6 that have occurred in FP7. The model used empirical information on the number of participants and the number of funded projects, together with data on project size (as measured by participant numbers) duration and average funding. Analysis of this information produced data on the functioning of, and relationships within, actual collaborative networks within the context of the Framework. Using this data in the model the research team was able to achieve a good match with the empirical data from EU-funded ICT networks in FP7: the model accurately reflected what actually happened. Because the simulation experiments using the empirical data closely match what actually happened, the model can be used as a test bed for potential policy choices.

References

Ahrweiler P, Pyka A, Gilbert N (2011) A new model for university-industry links in knowledge-based economies. J Prod Innovat Manag 28(2):218–235
Ahrweiler P, Gilbert N, Pyka A (2014a) Innovation policy modeling with SKIN. In: Johnston E et al (eds) Handbook of policy informatics. Routledge, London
European Commission (2009) Working document consultation on the future "EU 2020" strategy
European Commission/DG INFSO (2008) Panel report "Information Society Research and Innovation: delivering results with sustained impact – evaluation of the effectiveness of Information Society Research in the 6th framework programme 2003-2006", May 2008
Scholz R, Nokkala T, Ahrweiler P, Pyka A, Gilbert N (2010) The agent-based Nemo Model (SKEIN) – simulating European Framework Programmes. In: Ahrweiler P (ed) Innovation in complex social Systems. Routledge, London

Chapter 8
Towards a Prototype Policy Laboratory for Simulating Innovation Networks

Michel Schilperoord and Petra Ahrweiler

Abstract This paper presents an approach for designing and building a computational laboratory for research and innovation policy simulation, centred around the SKIN model. The aim of the paper is to bring together empirical and computational research for policy use. The SKIN model will be embedded in a workflow and an interfacing infrastructure that supports rich user interaction with the lab's simulation database.

8.1 Introduction

In the past few years, the SKIN model has been applied to various studies simulating the impacts of different policy choices on the structure, formation and effectiveness of research and innovation networks. This has answered an explicit demand of policymakers and evaluation experts in this area, who asked for policy impact simulation "to make evaluations relevant to policy options for intervention in RTD and innovation. Evaluations must relate observed parameters and impacts to the characteristics of the intervention. It must be possible to deduce what could and should be changed in the intervention to improve impacts. Accordingly, much more use should be made of ex-ante network analysis to simulate the impacts of intervention policy changes" (European Commission 2009: 18).

M. Schilperoord (✉)
UCD Innovation Research Unit (IRU), UCD Michael Smurfit Graduate Business School, University College Dublin, Carysfort Avenue, Blackrock, Dublin, Ireland
e-mail: michel.schilperoord@ucd.ie

P. Ahrweiler
EA European Academy of Technology and Innovation Assessment GmbH, Wilhelmstr. 56, 53474 Bad Neuenahr-Ahrweiler, Germany
e-mail: Petra.Ahrweiler@ea-aw.de

N. Gilbert et al. (eds.), *Simulating Knowledge Dynamics in Innovation Networks*, Understanding Complex Systems, DOI 10.1007/978-3-662-43508-3_8,

For example, working from datasets on EU-funded research networks in the area of information and communication technologies (ICT) (Ahrweiler et al. 2012, 2014b), the INFSO-SKIN application, specifically-developed for the Directorate General Information Society and Media of the European Commission (DG INFSO), helped to understand and manage the relationship between research funding and the goals of EU policy. The agents of the INFSO-SKIN application are research institutions such as universities, large diversified firms, small and medium-sized enterprises (SMEs). The model simulated real-world activity in which the Calls of the Commission specify the minimum number of partners, composition of partners, and the length of the project; the deadline for submission; a range of capabilities, a sufficient number of which must appear in an eligible proposal; and the number of projects that will be funded. The rules of interaction and decision implemented in the model corresponded to Framework Programme (FP) rules; to increase the usefulness for policy designers, the names of the rules corresponded closely to Framework Programme terminology. For the Calls 1–6 that had occurred in FP7, the model used empirical information on the number of participants and the number of funded projects, together with data on project size (as measured by participant numbers) duration and average funding. Analysis of this information produced data on the functioning of, and relationships within, actual collaborative networks within the context of the Framework Programme. Using this data in the model provided a good match with the empirical data from EU-funded ICT networks in FP7: the model accurately reflected what actually happened and could be used as a test bed for potential policy choices.

Altering elements of the model that equate to policy interventions such as the amount of funding, the size of consortia, or encouraging specific sections of the research community, enabled the use of SKIN as a tool for modelling and evaluating the results of specific interactions between policies, funding strategies and agents. Because changing parameters within the model is analogous to applying different policy options in the real world, the model could be used to examine the likely real-world effects of different policy options before they were implemented.

Another example is the application IPSE-SKIN (Ahrweiler 2012), which is currently developed in the context of national innovation policies (Dept. of the Taoiseach 2008) to answer the needs of the Irish economy to breed an innovation ecosystem with optimally structured university-industry-government networks. Informed by large new datasets coming from empirical research on the Irish national innovation system, with data on the regional, sector, university and firm level and on researchers and entrepreneurs, IPSE-SKIN is used to investigate the actors, designs, processes and policies of Irish innovation networks for developing and testing policy strategies for Irish innovation networks, and work on options for anticipating and analysing new developments to help the recovery of the economy. The high degree of empirical information in the model, together with the strong applied-orientation of experiments to examine how research networks change and adapt in response to different policy stimuli demonstrate the utility of this approach for policy makers.

In both examples, communication of results to the sponsors, clients and users of the studies was strongly supported by a graphical user interface that allowed policy makers to perform their own experiments without the requirement to directly change the code. This offered a kind of "lab experience", which was greatly appreciated from the user perspective.

However, the experiences from these applications and the confrontation with stakeholder and user demands, which became obvious during the studies, now ask for a more systematic approach to use the SKIN model for research and innovation policy impact simulation. We need to provide an integrated infrastructure of services and interfaces that can be offered to lab users. Our final aim is to provide lab facilities on the Internet by means of easy-to-use web apps with data visualisations designed for interaction with detailed data extracted from the lab's simulation database. How can this be realised, and what are the concrete steps to achieve this?

The conceptual set-up and the technical realisation must tackle both, internal issues of standardization, quality control, and effectiveness, as well as external features such as visualization and communication with and for lab users. The internal processes refer to the workflow of the lab and to the question how the SKIN model is exactly embedded in it.

In this paper, we present the computational designs developed so far for building a research and innovation policy laboratory *in silico* in order to inform policymakers about optimal network structures for research and innovation.

8.2 Realistic Simulation

The stakeholder-driven purpose of the lab requires to implement and test innovation policy or innovation management scenarios in a realistic and evidence-based way. The users of the lab are mostly not interested in results for scientific reasons, or wonder about the more general features and mechanisms of research and innovation networks. Their questions concern the SKIN application of their concrete network under investigation, which should therefore be informed by empirical data and be implemented with as much care to detail as necessary.

This often asks for handling large databases mapping the knowledge profiles and agent characteristics of research and innovation networks. Furthermore, adapting the SKIN model to various case studies of research and innovation networks requires qualitative analyses of decision contexts, agent behaviours and strategies and their implementation into the model. Only if the model features empirically observed processes leading to the empirically observed structures mimicking the causal mechanisms of real life, model-suggested interventions into these processes, in terms of empirical policy changes, can change empirical network structures in an expected way. Therefore, we are not only looking for SKIN reproducing empirical network structures but also reproducing the ways these are generated. The model needs to work with empirical information on both.

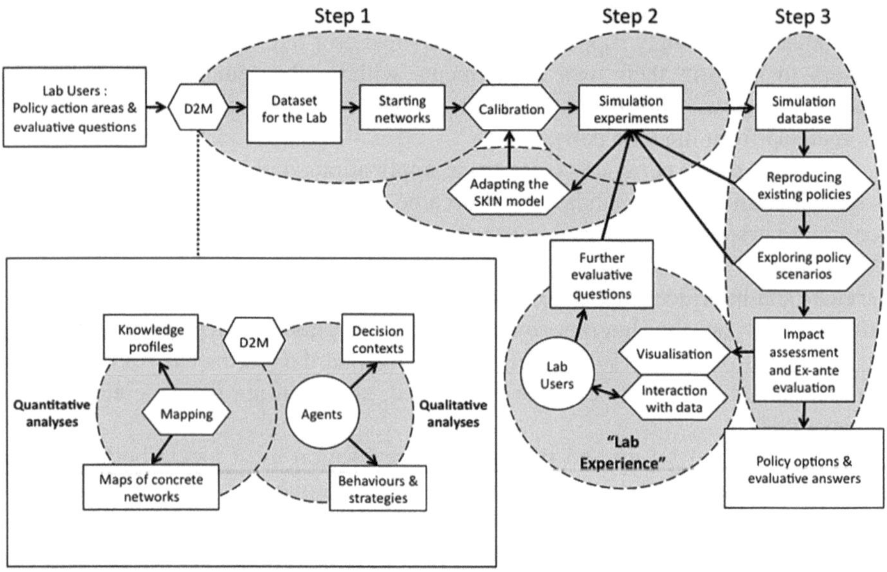

Fig. 8.1 Lab workflow and data-to-model (D2M) strategy

To find realistic starting configurations for simulating research and innovation networks under investigation, we need a close link between the model and this empirical information involving a high fraction of calibrated distributions concerning features of the agents and specific context conditions such as existing policies at work.

The lab prototype builds on a three-step data-to-model strategy (Fig. 8.1):

1. Use empirical data on research and innovation networks as data input to calibrate the SKIN model adapted to the study context ("starting networks")
2. Process this data in simulation experiments with different policy scenarios relevant for the lab users (mostly network monitoring, ex-post evaluation, impact assessment, ex-ante evaluation)
3. Analyse and interpret the simulated network data and provide results for lab users

8.3 Calibration

If we have empirical information as outlined available, we can apply our framework for transforming data points to model input deciding which categories of the databases can be used to set up the model.

While qualitative data will mostly inform the agent behaviours and the rules of the system, quantitative data can be used for calibrating the structure of the research

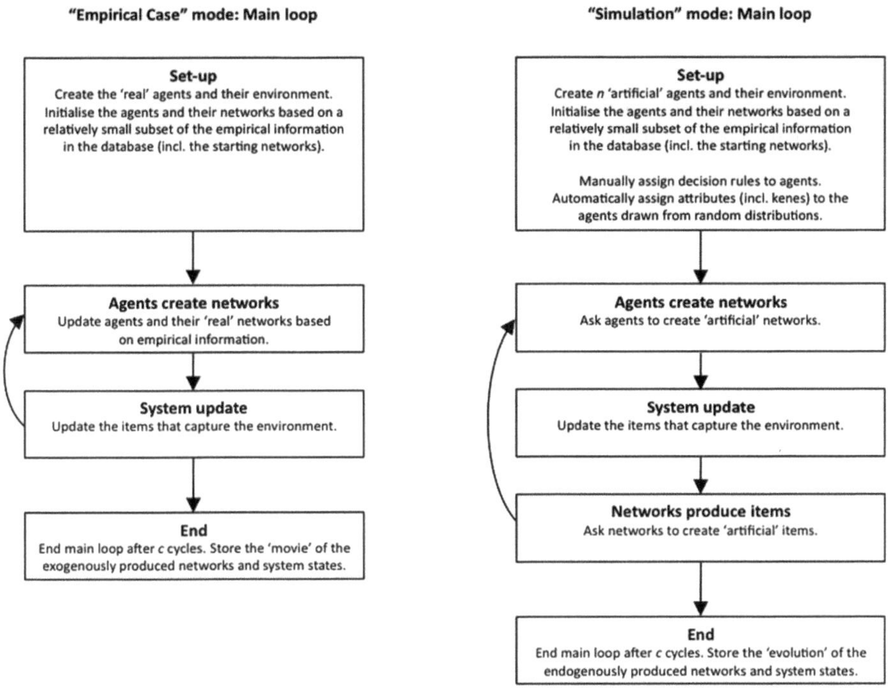

Fig. 8.2 Calibration: empirical case mode vs. simulation mode

and innovation network under investigation. Our calibration procedure tries to match the structure of the 'real', empirically observed, research and innovation network with the structure of the 'artificial' network generated by the simulation model, before we even start with any simulations, i.e. we just try "to reproduce the database" with the model. The database is connected to the model using MySQL, which means that data can be read directly from within the model with standard SQL queries (the interface platform has a switch called "empirical case" to do this).

Figure 8.2 shows how the main loop differs when running the model in empirical case mode (EMP). In EMP-mode agents are 'real' in the sense that they directly point to the network participants stored in the database (e.g. company X or university Y). In set-up, a relatively small subset of the information on starting networks in the database is used to create and initialise 'real' networks between these agents. These networks are 'real' in the sense that they directly point to the networks stored in the database (e.g. partnership Z). After set-up, each cycle of running the model amounts to updating the agents and the networks based on empirical information in the database. This stepping through the cycles is similar to playing back the "movie" of the agents creating these networks. The beauty of this approach is that it allows us to use in EMP-mode many of the procedures that are also used when running the model in simulation mode, e.g. the procedures for system updates (each cycle) and storing simulation outputs (at the end of the last cycle). This makes calibration much easier.

The calibration procedure is:

1. Switch the platform into 'empirical case' mode.
2. Set-up the case. This starts the initialisation of the model using the empirical database. Agents and environment are created based on the information in the database.
3. Start the SKIN application. The number of iterations corresponds to the time covered in the database (if time is recorded).
4. Each iteration, the platform updates its internal representation of the network of the 'real' agents, which are all directly linked to information on actual organisations in the database. The updating process uses the empirical information retrieved from the database. This stepping process (not simulation) recreates the evolution of the 'real' research and innovation network and, taking advantage of the platform's user interface and analytical features, we can observe how this network structure has evolved.
5. Switch the platform into 'simulation' mode.
6. Set-up the case. This starts the initialisation of the model using input parameters extracted from the empirical database. Agents and environment are automatically created based on these parameters. (In 'simulation' mode, some items might not be created at initialisation, since they will be outputs of the simulation)
7. Start the SKIN application.
8. Each iteration, the platform updates its internal representation of the network of the agents, based on the endogenised procedures (which might be calibrated by qualitative data, see above), i.e. created by the agents and outcome of the policy strategies.

The 'real' and 'artificial' networks, automatically stored for each iteration in series of data files (GML, GraphML and GEXF), can be compared using the complete set of network analysis tools, either using the algorithms that are built into the platform, or from software libraries like Gephi (Bastian et al. 2009), NetworkX, or other. If we are confident that we have reproduced the major stylised facts of the research and innovation networks under investigation that have been extracted from empirical information, the model can be used as a laboratory to change the design of instruments and to analyse divergent developments.[1]

[1] During the course of the INFSO-SKIN study where we worked with a large database, which was pre-structured into six Calls of FP7, the study team (Nigel Gilbert, Andreas Pyka, Michel Schilperoord, Petra Ahrweiler) started to develop a "multi-calibration/validation" strategy, where the model was supposed to match the network configuration at the end of each Call and start again with the network configuration of the next. This not only gave six points for assessing

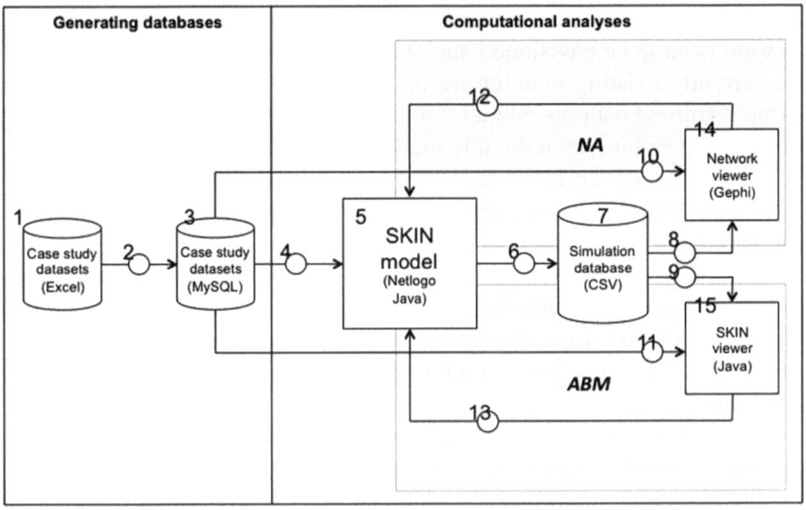

Fig. 8.3 General workflow (middleware: 2, 4, 6, 8, 9, 10, 11, 12 and 13)

8.4 Workflow

Workflow, meaning the often tedious interlinking of resources, is an important topic when it concerns empirical calibration and validation. Assessing the performance of models against observed data requires a constant review of the entire protocol for techniques used in constructing and interrogating (i.e. experimenting with) these models (Windrum et al. 2007; Lorscheid et al. 2012). The ABM community is to respond with standard techniques: workflows and protocols that are tailored to commonly used resources and techniques for constructing ABMs and integrate these with widely used ABM tools such as NetLogo and RePast. While standard techniques can be helpful, reality is that agent-based modellers will often need to adapt their workflows and protocols to very particular problems concerning their resources and techniques. So they also need to be creative. But how can one match and reconcile both needs: robustness and creativity?

Figure 8.3 displays an open adaptable software platform with a key role for the SKIN model surrounded by middleware—the software that glues things together. This platform was developed during the studies mentioned above and already applied and tested within these applications. It will be further elaborated and extended within current and future research. To develop our calibration and

the calibration/validation performance instead of the usual one, but also provided an opportunity for model learning and fine-tuning. This strategy can be always used when empirical databases are divided into sections containing enough information about time and interim states. The idea could not be followed up systematically at that time, but seems worthwhile to pursue in future work.

validation methodologies, more ways of interlinking component workflows with middleware need to be envisioned and tested.

The currently existing middleware is designed to (1) interlink the lab components (the empirical datasets, models, an integrated simulation database, analytical tools and visualisations) in a flexible manner, (2) deal with data-related problems, (3) facilitate a general workflow that is explorative and data-driven, and (4) exploit the ABM's strengths by covering an as wide as possible range of use cases, including *unforeseen* use cases.

For calibrating the SKIN model with empirical data on research and innovation networks, we developed software for bridging between databases and the SKIN model. This involved formatting the network data in such a way that it can be read into the SKIN model in order to structure the initial network of agents at the start of the simulation.

The middleware links displayed in Fig. 8.3 support different parts and stages of the general workflow: Middleware 2 and 4 provide support for the data-to-model process: storing and protecting the datasets (e.g. Excel tables), conversion to MySQL, cleaning of datasets, using the MySQL database and connecting the database to the SKIN model.

Middleware 6 supports the simulation process: using the SKIN model (Netlogo or Java version, depending on total computational load estimated for experiments), writing the SKIN model results to the simulation database, and building the simulation database and administration thereof. Middleware 8 and 9 support the visualisation of simulation results with Gephi and exploration of simulation results with the SKIN viewer. The latter is an existing tool developed in Java for the SKIN model for browsing the (large) integrated simulation database and comparing results with experiments.

Middleware 10 and 11 make it possible to comparing empirical data and simulation results with Gephi and the SKIN viewer. A core set of indicators is computed on both sources of data, in direct support of calibration and validation procedures. Yet simulation results allow for an additional set of indicators to be computed that cannot be measured in empirical studies. This provides valuable extra information for understanding the causal processes and mechanisms creating the structures and states captured with the core indicators set.

Middleware 12 and 13 provide important feedback from the simulation results to choice of SKIN model parameters and settings. This is crucial not only for calibration and validation of the SKIN model (using Gephi and the SKIN Viewer) but also, as mentioned above, for systematising and reviewing the entire experimentation protocol.

The visualisation and evaluation of the results stored in the simulation database is helped by the SKIN Viewer tool, which was programmed in Java for the INFSO-SKIN application. It provides several functions to visualise and compare the large volumes of data generated in experiments with the INFSO-SKIN model, including a t-test for statistically significance of findings. It can be easily adapted to other SKIN applications (Fig. 8.4).

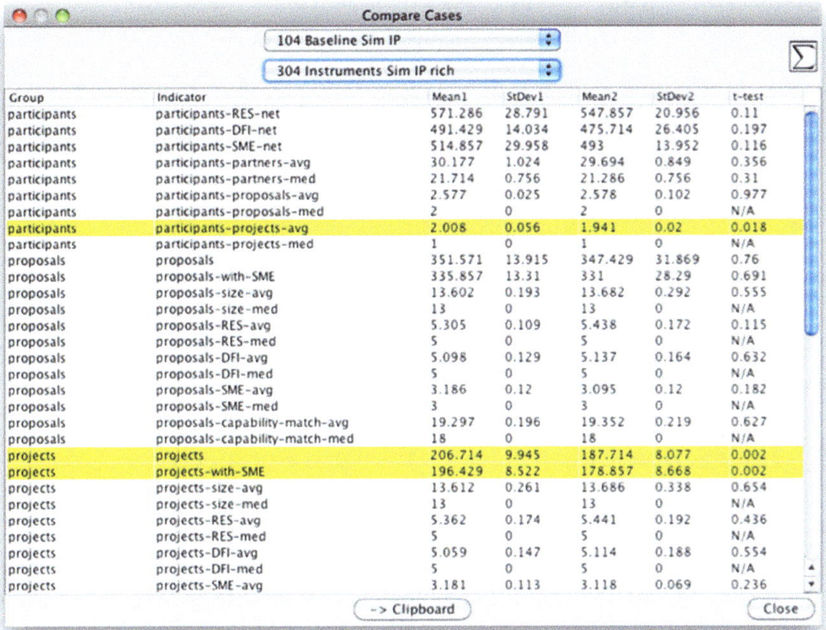

Group	Indicator	Mean1	StDev1	Mean2	StDev2	t-test
participants	participants-RES-net	571.286	28.791	547.857	20.956	0.11
participants	participants-DFI-net	491.429	14.034	475.714	26.405	0.197
participants	participants-SME-net	514.857	29.958	493	13.952	0.116
participants	participants-partners-avg	30.177	1.024	29.694	0.849	0.356
participants	participants-partners-med	21.714	0.756	21.286	0.756	0.31
participants	participants-proposals-avg	2.577	0.025	2.578	0.102	0.977
participants	participants-proposals-med	2	0	2	0	N/A
participants	participants-projects-avg	2.008	0.056	1.941	0.02	0.018
participants	participants-projects-med	1	0	1	0	N/A
proposals	proposals	351.571	13.915	347.429	31.869	0.76
proposals	proposals-with-SME	335.857	13.31	331	28.29	0.691
proposals	proposals-size-avg	13.602	0.193	13.682	0.292	0.555
proposals	proposals-size-med	13	0	13	0	N/A
proposals	proposals-RES-avg	5.305	0.109	5.438	0.172	0.115
proposals	proposals-RES-med	5	0	5	0	N/A
proposals	proposals-DFI-avg	5.098	0.129	5.137	0.164	0.632
proposals	proposals-DFI-med	5	0	5	0	N/A
proposals	proposals-SME-avg	3.186	0.12	3.095	0.12	0.182
proposals	proposals-SME-med	3	0	3	0	N/A
proposals	proposals-capability-match-avg	19.297	0.196	19.352	0.219	0.627
proposals	proposals-capability-match-med	18	0	18	0	N/A
projects	projects	206.714	9.945	187.714	8.077	0.002
projects	projects-with-SME	196.429	8.522	178.857	8.668	0.002
projects	projects-size-avg	13.612	0.261	13.686	0.338	0.654
projects	projects-size-med	13	0	13	0	N/A
projects	projects-RES-avg	5.362	0.174	5.441	0.192	0.436
projects	projects-RES-med	5	0	5	0	N/A
projects	projects-DFI-avg	5.059	0.147	5.114	0.188	0.554
projects	projects-DFI-med	5	0	5	0	N/A
projects	projects-SME-avg	3.181	0.113	3.118	0.069	0.236

Fig. 8.4 Screen shot of the SKIN Viewer tool—summary table with t-test results

Obviously, the lab's main component workflows centre round network analysis (NA) and agent-based modelling with SKIN. NA casts a perspective on structures and states of innovation networks and ABM adds a perspective on causal processes and mechanisms creating these structures and states (cf. Ahrweiler 2010). In practice, these two workflows feed each other. For example, the empirically observed network data can be used as starting configurations for setting up a SKIN simulation (through middleware depicted as 12 in Fig. 8.3). Inversely, the SKIN model can be used to create artificial network data (through middleware depicted as 8 in Fig. 8.3).

The second issue the conceptual set-up and the technical realisation of the lab prototype must be able to deal with lab-external features such as visualisation and communication with and for lab users.

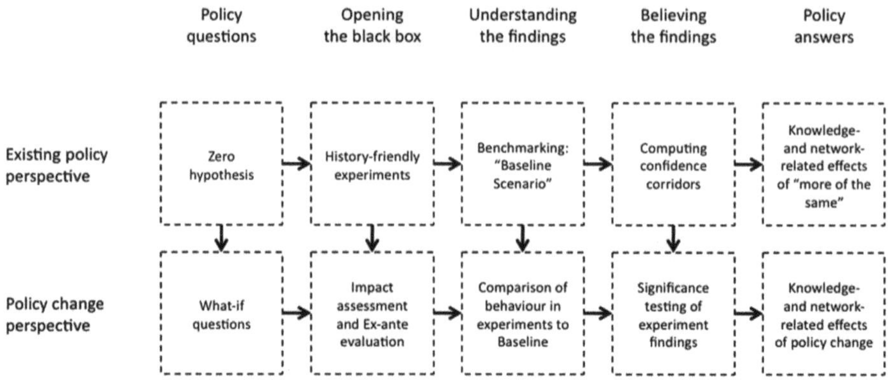

Fig. 8.5 The lab functions from a user's perspective

8.5 Interaction with the Lab: A User's Perspective

For a lab prototype on research and innovation policy we have to envisage that we will have to deal with simulation data from case studies that differ importantly in focus, scope and scale, yet have a common conceptual reference: policy questions concerning research and innovation networks (Fig. 8.5).

From a user's perspective, there are three issues, where the first and the third are yet outside the scope of themes we already discussed, but which are central to the lab user if not the only issues any user is interested in:

(i) The policy-relevant questions the user wants the simulation to answer
(ii) The credibility of the findings supported by the evidence base of the realistic simulations calibrated by empirical data (often provided by the users themselves)
(iii) Understanding the answers the simulation findings provide

8.5.1 *Evaluative Questions*

Depending on whether the workflow is applied to a tender study with a clear client demand behind it, or to a more open research project sponsored by a general funding agency, the questions the simulation needs to answer will be more or less pre-defined from the onset of a project.

However, we learnt that even in very client-driven studies, enough time should be dedicated to identifying and discussing the exact set of questions the stakeholders of the work want to see addressed. We found that the best way to do this is applying an iterative process of communication between study team and clients, where stakeholders learn about the scope and applicability of the methods, and where researchers get acquainted with the problems policy makers have to solve and with the kind of decisions, for which sound background information is needed. This iterative process will result in an agreed set of questions for the simulation, which will very often decisively differ from the set proposed at the start of the study. For the INFSO-SKIN study, a so-called "Steering Committee" was assigned to us consisting of policy makers and evaluation experts of DG INFSO. For the IPSE-SKIN application, we have employed a Stakeholder Board of high-profile Irish innovation policy representatives.

The strength of our type of methodology lies in the opportunity to ask what-if questions (ex-ante evaluation), an option, which is normally not easily available in the policy-making world. To find the right set of what-if questions, the benchmark question to go from is the "zero hypothesis": what if there are zero changes? This just follows the extension of the time horizon for the simulated research and innovation networks without changing the rules (the so-called Baseline Scenario). To do this, just "more of the same" is made available to the agents and the networks. Answering the "zero hypothesis" question is important for two reasons: First, the original line of development and network evolution is extended and the sustainability of the network architectures is tested. Second, we can use this scenario as a kind of benchmark and compare its outcomes with the results of further experiments, as naturally no empirical data about the future are available for comparisons.

If the networks created by real life and those created by the agent-based model qualitatively correspond closely, the simulation experiments can be characterized as *history-friendly* experiments, which reproduce the empirical data and cover the decisive mechanisms and resulting dynamics of the real networks: 'History-friendly models are formal models which aim to capture—in stylized form—qualitative theories about mechanisms and factors affecting industry evolution, technological advance and institutional change put forth by empirical research in industrial organization, in business organization and strategy, and in the histories of industries. They present empirical evidence and suggest powerful explanations. Usually these "histories" (. . .) are so rich and complex that only a simulation model can capture (at least in part) the substance, above all when verbal explanations imply non-linear dynamics' (Malerba et al. 1999: 3–4).

This qualifies the model to be applied for *in silico* experiments, which take these outcomes as a starting point and test the potential future developments in the sense of the *ex-ante evaluation* outlined above. INFO-SKIN, for example, carried out a number of experiments to simulate policy options in Horizon 2020 and their consequences for the future ICT research landscape in Europe. The results were compared with the baseline scenario of the unmodified FP7 ICT Calls 1–6. Differences between the behaviour in the experiments and that in the baseline could be attributed to the effects of the policy change.

What types of questions are possible? As SKIN is about knowledge dynamics and inter-organisational network dynamics, evaluative questions can address effects of policy changes on two levels—knowledge and agents/networks.

The knowledge-related indicators (cf. Ahrweiler et al. 2011) demonstrate the additional advantage of a simulation to record the development of an "un-observable". Empirically, we cannot see and only with difficulty can we measure knowledge generation, knowledge exchange, and learning. In the simulation, the effects of knowledge exchange processes are measured as knowledge flows between the agents in the networks. With SKIN, we can see what policy changes do to the amount of knowledge in the system and the characteristics of the knowledge flows.

The network-related indicators show many statistical properties of networks that are standard applied in network analysis, thus allowing the same type of inquiry. Simulation offers the advantage of seeing effects of policy changes over many (new) instantiations of these networks, and therefore gives us information on their dynamic characteristics, e.g. are networks adaptive to changing conditions, robust, sustainable? Which agents are they attracting? Are they productive? (not only in terms of knowledge production but also, for example, in terms of creating niches within the larger network structure for accelerated growth). For the latter type of questions, the simulation's main loop (Fig. 8.2) offers a facility—the "networks produce items" step—that can be adapted to particular applications.

Evaluative questions can address both, the knowledge and the network level. For example, the agreed set of evaluative questions for the INFSO-SKIN application only contained one question for the knowledge level (the first one) and various questions for the agents/networks level. These were:

- What if there are changes to the thematic areas, which are funded?
- What happens if there are changes in the funding instruments (small actions, large actions)?
- What if there are interventions concerning the scope/outreach of funding?
- What happens if there are changes in the funding level (increase, decrease)?
- What if there are interventions concerning the participation of certain actors in the network (e.g. SMEs)?

Also for the IPSE-SKIN application, questions concern structure and operation of networks, their appropriate scale and scope (present technology/industry sectors and new fields) and how the networks are adapted and embedded in context of national, regional and local agendas. Evaluation concentrates on finding the optimal network structures for: (1) supporting enterprises and promoting competitiveness, (2) strengthening the whole innovation cycle (entrepreneurship, innovation, commercialisation), (3) stimulating the green sector (energy security, protecting the environment) and (4) improving the efficiency and quality of public services and regulation (Building Ireland's Smart Economy 2008: 39, reference in bibliography).

8.5.2 Helping Users to Understand and Interpret the Findings

Our user's perspective, geared towards asking evaluative questions, puts heavy demands on the lab's interfaces system, not only for being able to adapt the lab to specific questions, but also for allowing users to understand and trust the findings. We already (in Sect. 8.4) pointed out middleware as a flexible way of adapting the lab and setting up a workflow that is generally robust. Our experience is that *interaction by users with data* from network analysis (NA) and simulations (ABM) can also help creating a more solid basis for understanding and trust, adding to the traditional technical style of reporting findings.

This challenge is illustrated by access to the lab's simulation database. We find the richness of this database (the INFSO-SKIN database contains many gigabytes of disaggregated data) valuable in our lab approach. We also recognize however, that the sheer amount of new NA/simulation data can present limitations for understanding the patterns and information it contains. The good news is that sophisticated ways of interfacing and interacting with NA/simulation data are increasingly feasible. The new possibilities are reflected and foreshadowed by the numerous extensions created by NA/ABM communities, as illustrated by growing number of Netlogo extensions and Gephi plugins. They can build on technical developments in providing data-driven applications (or apps) on the Internet with strong visualisations. Possibilities are illustrated by MIT's Observatory of Economic Complexity (Simoes and Hidalgo 2011; Hausmann et al. 2011).

To optimally benefit from these possibilities, the lab provides four interfaces:

- The users interface (UI) for remote access to the lab,
- The web applications interface (API) for bringing to life the lab's data,
- The modelling processes interface (MDI), including the middleware, for setting up and controlling computational processes, and
- The databases interface (DBI) for storage of large streams of data feeding in and out of the lab.

The UI facilitates remote access to the lab's web pages using desktop or mobile devices with a standard web browser. It connects to the API for quickly serving up web pages containing apps that enable visualisation of, and interaction with, the lab's data. Expert users of the lab use the MDI for setting up and controlling the lab's computational processes, whereas empirical research teams have their own interface for binding their data inputs to the lab. The DBI also offers tools for inspecting and managing the lab's results, which are automatically stored in a single, integrated simulation database.

The computational laboratory *in silico* can inform policy and management on optimal network structures for innovation performance adapted to contextual conditions. Our simulations will serve as a laboratory to experiment with social life in a way that we cannot do empirically due to methodological reasons (cf. Ahrweiler and Gilbert 2005). Using this tool, we can understand innovation dynamics in complex social systems and find their potential for design, intervention and control.

Acknowledgment We gratefully acknowledge the financial support from the Irish Government's Programme for Research in Third Level Institutions (PRTLI 5) project grant entitled 'Innovation Policy Simulation for the Smart Economy (IPSE)'.

References

Ahrweiler P (ed) (2010) Innovation in complex social systems. Routledge, London

Ahrweiler P (2012) Conceptual framework of computational policy research: a guiding document and reference for the IPSE consortium. Deliverable D1.1, Project report, http://www.ipse.ie

Ahrweiler P, Gilbert N (2005) Caffè nero: the evaluation of social simulation. J Artif Soc Soc Simulat 8(4):14, http://jasss.soc.surrey.ac.uk/8/4/14.html

Ahrweiler P, Pyka A, Gilbert N (2011) A new model for university-industry links in knowledge-based economies. J Prod Innovat Manag 28:218–235

Ahrweiler P, Schilperoord M, Gilbert N, Pyka A (2012) Final report – using network analysis to monitor and track effects resulting from changes in policy intervention and instruments. SMART 2010-0025, DG INFSO, January 2012, Brussels

Ahrweiler P, Gilbert N, Pyka A (2014b) Innovation policy modeling with SKIN. In: Johnston E et al (eds) Handbook of policy informatics. Routledge, London

Bastian M, Heymann S, Jacomy M (2009) Gephi: an open source software for exploring and manipulating networks. In: International AAAI conference on weblogs and social media

Department of the Taoiseach (2008) Building Ireland's smart economy: a framework for sustainable economic renewal. Dublin

European Commission (2009) Using network analysis to assess systemic impacts of research. Workshop report, DG INFSO, Brussels, March 2009

Hausmann R, Hidalgo CA, Bustos S, Coscia M, Chung S, Jimenez J, Simoes A, Yildirim M (2011) The atlas of economic complexity. Puritan, Cambridge, MA

Lorscheid I, Heine B-O, Meyer M (2012) Opening the 'black box' of simulations: increased transparency and effective communication through the systematic design of experiments. Comput Math Organ Theor 18(1):22–62

Malerba F, Nelson R, Orsenigo L, Winter S (1999) 'History-friendly' models of industry evolution: the computer industry. Ind Corp Change 8:3–40

Simoes AJG, Hidalgo CA (2011) The Economic Complexity Observatory: an analytical tool for understanding the dynamics of economic development. In: Workshops at the twenty-fifth AAAI conference on artificial intelligence, http://atlas.media.mit.edu

Windrum P, Fagiolo G, Moneta A (2007) Empirical validation of agent-based models: alternatives and prospects. J Artif Soc Soc Simulat 10(2):8, http://jasss.soc.surrey.ac.uk/10/2/8.html

Part III
Applying SKIN to Innovation Sectors

Chapter 9
Modelling the Emergence of a General Purpose Technology from a Knowledge Based Perspective: The Case of Nanotechnology

Benjamin Schrempf and Petra Ahrweiler

Abstract Nanotechnology, the manipulation and control of matter at the scale 1–100 nm, proves to have an increasing socio-economic impact on its way to become the key-technology of the twenty-first century. It has already found applications in various industrial sectors such as information and communication technology, pharmaceuticals, materials and manufacturing, or biotechnology. Nanotech is a so-called "General Purpose Technology (GPT)": with a broad range of applicability and spread in many industries, its innovation networks considerably differ from those of other emerging technologies. Having a strong semi-conductor, materials, and biotechnology industry, the highest "Revealed Technological Advantage" in nano-technology of Western Europe, and a high share of nanotechnology patents, Ireland seems to be a promising case study for investigating the dynamics of nanotechnology knowledge, its role in the economy, and the effect of policies on both.

In this paper, we address the specific characteristics of GPT innovation networks and suggest ways to model them using Ireland as an empirical case study. We discuss literature providing important stylised facts about nanotechnology/GPT and suggest how they can be implemented into a SKIN application simulating Irish nanotech innovation networks.

9.1 Introduction

Nanotechnology as scientific field already has a strong impact on the development of the economy and will have even more so in the future, making it 'the key-technology of the twenty-first century' (Bhattacharya and Shilpa 2011).

B. Schrempf (✉) • P. Ahrweiler
EA European Academy of Technology and Innovation Assessment GmbH, Wilhelmstr. 56, 53474 Bad Neuenahr-Ahrweiler, Germany
e-mail: Benjamin.Schrempf@ea-aw.de; Petra.Ahrweiler@ea-aw.de

N. Gilbert et al. (eds.), *Simulating Knowledge Dynamics in Innovation Networks*, Understanding Complex Systems, DOI 10.1007/978-3-662-43508-3_9,

It is often defined as the manipulation and control of matter at the scale 1–100 nm with applications to be found in numerous sectors such as the information and communication technology industry (ICT), pharmaceutical industry, material and manufacturing industry or biotechnology.

The analysis of nano-related research and innovation, science and technology, has been of huge interest in recent years. Specific characteristics of nanotechnology were investigated, for instance concerning the generation and diffusion of knowledge, the scope, structure and dynamics of the field, its interdisciplinarity, the management of nanotechnology research and innovation as well as its impact on the economy. With nanotechnology being widely regarded as a general purpose technology (e.g. Teichert 2012), nanotechnology patterns differ in many aspects from other emerging technologies in not showing a comparable broad range of applicability. As knowledge is one major prerequisite for innovation, the investigation of its generation, structure and dynamics are of great importance. Taking a knowledge perspective on the emergence of nanotechnology therefore seems to be appropriate in order to understand how nanotechnology may evolve thereby influencing the economy. The insights gained can help policy makers designing efficient policy measures for fostering knowledge generation and commercialisation, thus generating economic output.

In this contribution, we describe a number of stylised facts from literature and empirics on the emergence of general purpose technologies (GPTs) in general and nanotechnology in particular. A number of indicators characterising a GPT are identified and applied to the SKIN model. This is to investigate whether GPTs can already be found emerging within the simulation model. With our work we contribute to the literature of GPTs by modelling the emergence and diffusion of GPTs within networks from a knowledge perspective.

The empirical data used is based on a case study of the Irish nanotechnology sector. With a strong semiconductor sector, materials sector and biotechnology sector (Forfás 2010), the highest revealed technological advantage in nanotechnology in western Europe (OECD 2009) a rapid growth rate and a high share of nanotechnology patents, Ireland provides a promising sample to study the dynamics of nanotechnology knowledge, its role in the economy, and the effect of policy measures on nanotechnology evolution.

The remainder of the paper is organized as follows. Section 9.2 will give a short introduction to the literature of general purpose technologies and existing models. Subsequently, it is discussed whether nanotechnology can be considered a GPT and why we need distinct GPT models. Section 9.3 identifies stylised facts a meaningful GPT model should be able to replicate. Looking at several different indicators the SKIN model is examined with respect to two of the most important characteristics. Additional measures for the GPT features remaining are introduced and proposals for an improved knowledge and value chain representation are given, Sect. 9.3 concludes.

9.2 General Purpose Technologies

9.2.1 GPT Literature

The term general purpose technology was coined by Bresnahan and Trajtenberg (1995) who define GPTs as 'characterized by pervasiveness, inherent potential for technical improvements and 'innovational complementaries'.' (p. 83). GPTs are also said to play a major role in economic growth: by incorporating them in economic models it is possible to explain growth endogenously and provide an integrated model of growth and business cycles (Bresnahan and Trajtenberg 1995). In literature, the term 'enabling technologies' is often used similarly and refers to the character of GPTs: combined with other technologies of the application sectors (ASs), the emergence of GPTs triggers innovations (Bresnahan 2010).

According to Schiess (2011) the literature on GPTs can be classified in two main strands: The first strand mainly deals with the invention of GPTs, the second strand with their diffusion. Models focussing on the invention of GPTs can be further divided into models that emphasize the connection between R&D efforts and growth triggered by the emergence of one or more GPTs, and models that underline the need for complementary innovations in order to make use of the GPTs. Both strands rather focus on the explanation of long-term growth whereas lifecycles (of both, the GPT as well as the application technologies) are not taken into account. In the complementary-focussed models the diffusion of a GPT is determined significantly by the complexity of the complementaries to be developed for the application of the technology (Bresnahan 2010) and might differ from AS to AS.

Literature on the diffusion of GPTs either deals with domestic diffusion or sectoral diffusion characteristics, thereby focussing on the s-curve shaped pattern of diffusion. Other models apply a North/South framework, emphasising relative productivity differences. Most of these models mainly try to explain the productivity paradox—the empirical fact that productivity slows down in the early phase of the advent of a new GPT. This phenomenon was first described by Robert Solow (1987), referring to the developments after the introduction of computers in the economy, which did not lead to higher productivity. However, several years after the introduction, productivity and growth rates rose again. The same effects could be identified for other GPTs. These models are therefore attempting to explain more medium to long-term waves of growth (as fist described by Kontratjeff) by the emergence of one or more GPTs (Bresnahan 2010; Schiess 2011).

The models not only differ in endogenous emergence of the GPT or an exogenous GPT shock, but also in how GPTs are replaced. Some use Schumpeterian competition, modelling new GPTs as replacing the old one and never allowing for more than one GPT being used at the same time. Others allow for more than one GPT affecting the application sectors at the same time (van Zon et al. 2003; Schiess 2011).

Both main bodies of literature—the first dealing with the emergence of GPTs as well as the second, focussing on the diffusion of GPTs—ignore however the role of networks in either of the processes. Carlaw and Lipsey (2006) provide a central model focussing on the invention of GPTs and the importance of R&D efforts. The authors model three sectors (fundamental research, applied research, consumer goods), each of which has its specific production function. Continuous growth in the model is only possible through the repeated introduction of a new GPT. Uncertainty exists concerning the effects of R&D activities on the sequence of GPT arrivals and effect they unfold. However, the model also has some shortcomings: it keeps the emergence of GPTs exogenous, it does not model the different effects of the GPT on different sectors and it does not model the emergence and diffusion of the new GPT knowledge within the system. With a network approach in which innovations emerge from the collaboration of actors by combining their knowledge, GPT emergence could be endogenised.

Aghion and Howitt (e.g. 1998) are able to replicate the s-shaped curve of diffusion, at the same being able to explain the productivity paradox taking into account time lags and critical masses of users of a GPT needed to unfold a GPT's full effect (Schiess 2011). However, these models are ignoring the importance of networks and different network settings within and between the application sectors, the effect a GPT has on the industry architecture of a sector, and possible backward, forward and spillover effects. With our approach, we are bringing together the broad literature of GPT emergence and diffusion, and the network-focussed SKIN model.

9.2.2 Nanotechnology as General Purpose Technology

With the literature on GPTs outlined, the question arises which technologies are GPTs. Undisputed examples for GPTs can be found in Computers (Helpman and Trajtenberg 1994) and the steam engine (Lipsey et al. 2005). More disputed examples include the laser technology or biotechnology (Lipsey et al. 1998). Several studies have attempted to identify Nanotechnology as a GPT, of which two should be named here explicitly:

Based on the definition of Bresnahan and Trajtenberg (1995) for a technology to be a GPT it has to show pervasiveness, inherent potential for technical improvements and innovational complementaries, Youtie et al. (2007) conduct an analysis of patents and patent citations, concluding that nanotechnology show these characteristics and may therefore be considered to be a GPT. Teichert (2012) also draws from this definition of GPTs and analyses patents and publications on the level of growth rates, diffusion and citations, also underlining that there is strong evidence for nanotechnology to be a GPT.

The strong evidence given in the literature on the GPT character of nanotechnology therefore can justify the use of nanotechnology data to validate and later calibrate a GPT version of the SKIN model.

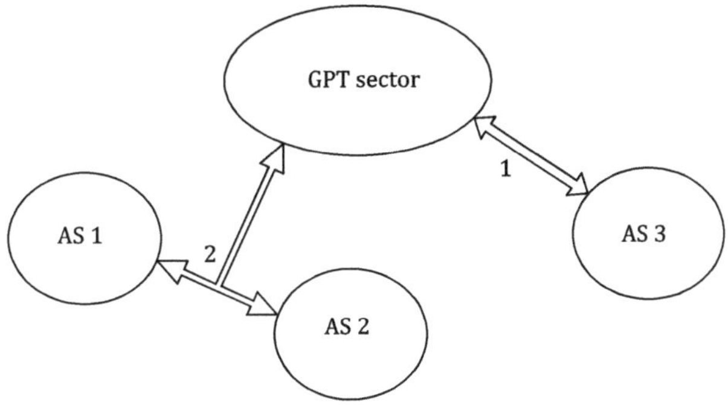

Fig. 9.1 Knowledge combination between different areas

9.2.3 The Uniqueness of General Purpose Technologies

A separate simulation effort for GPTs requires these types of technology to have decisive different features from other, 'standard' technologies. GPTs are technologies that are used in a much broader range of sectors than other technologies, they are evolving themselves by innovation and at the same time requiring many other complementary innovations, stemming from multiple other sectors. It is the *combination* of these three characteristics (pervasiveness, inherent potential and innovational complementaries) which make GPTs a proper field of research.

From a knowledge perspective, it seems also reasonable to treat GPTs differently from other technologies. Due to its interdisciplinarity (Porter and Youtie 2009), nanotechnology as a GPT merges formerly unconnected or only weakly connected disciplines. In this respect aspects of cognitive proximity, cross-fertilisation and knowledge complementarity are playing an important role from a GPT perspective (Teichert 2012). The combination of previously uncombined technologies in turn may lead to the broad applicability (Nikulainen 2010) or generality of nanotechnology, which can for instance be applied in different sectors such as biotechnology and ICT. According to Bresnahan and Trajtenberg (1995), the application is characterised by two different types of externalities. Within the AS, nanotechnology knowledge might trigger innovations (e.g. nano-based pharmaceuticals) by combining the nanotechnology knowledge with the knowledge of the AS (vertical externalities; see 1 in Fig. 9.1) based on complementariness (e.g. nano-enabled bio-sensors). Between the ASs, the GPT can enable the combination of knowledge from formerly unconnected fields due to its generality (horizontal externalities; see 2 in Fig. 9.1).

These trajectories of the emerging GPT knowledge on the one hand and the merging effect it may have on the other hand, should be reflected in the structure of the knowledge network in the SKIN model.

As for the collaboration between firms, GPT innovation networks can be expected to show similar characteristics to the networks of knowledge, which is creating new cross-sectoral connections not only directly between the ASs (in step 3 and 4 in the value chain depicted in Fig. 9.2) but also through firms and sub-networks specialised on the GPT knowledge (step 1 and 2, Fig. 9.2).

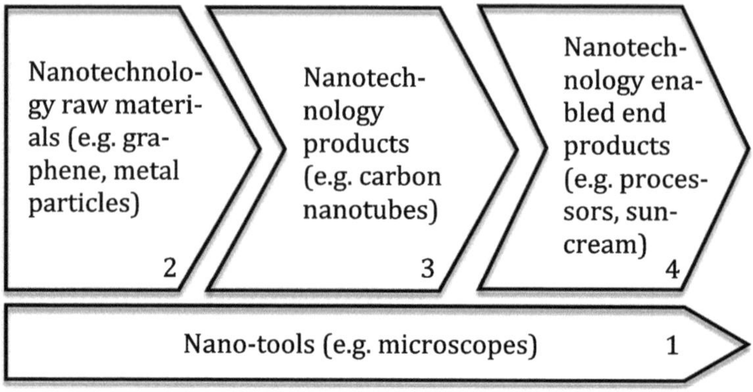

Fig. 9.2 Nanotechnology value chain (Forfás 2010)

Thus GPT networks, i.e. networks of firms specialised in the GPT may not only have their own characteristics and evolutionary trajectories, they also connect networks with different structures increasingly down the value chain where nano-enabled products are applied, thereby influencing their evolution and getting influenced by them. As for the knowledge perspective, from the innovation and collaboration perspective, it again seems to be reasonable to treat GPT innovation networks differently than other innovation networks.

Policy measures for GPTs should also differ from 'standard' policy measures for several reasons. Genet et al. (2012) studied similarities of technology transfer patterns in different industries, comparing nanotechnology with biotechnology and microelectronics. They found that technology transfer patterns in the nano sector are more similar to the patterns in microelectronics than to the patterns in biotechnology. Also, the knowledge transfer between large firms and research institutes seems to play a bigger role in nano-electronics than in nano-biotechnology. Therefore they argue, the implementation of policy measures from biotech in nanotechnology would be inappropriate (Genet et al. 2012). It seems reasonable to argue that nanotechnology policy measures should take into account the characteristics of nanotechnology research and innovation in general and the R&D patterns specific to the respective R&D sector in particular. Another example can be found in the discussion about nanotechnology safety, which plays a decisive role in nano-medicine and nano-food, but is almost non-existent in nano-electronics.

9.3 Modelling GPT Networks from a Knowledge Perspective: Identifying GPTs in SKIN

9.3.1 *Stylised Facts of Nanotechnology as a GPT*

In the following, several stylised facts mainly drawn from the literature review and the definition of GPTs are described and applied to the standard SKIN model. This will show whether and to what extent the existing basic SKIN model is already capable to replicate first these stylised facts and second some results from the Irish case study.

Investigating the knowledge level of the SKIN model, we look at two of the most important characteristics of GPTs and draw some hypotheses on how these characteristics should be reflected on the knowledge level in the SKIN model. The indicators proposed investigate how knowledge evolves, spreads throughout the network and how the different pieces of knowledge are related to each other. As Hall and Trajtenberg (2004) state, to identify GPTs and confine them from non-GPTs, it is necessary to take a multivariate approach since looking only at one indicator 'may be misleading' (p. 7). Again, we are basing the identification of indicators on the definition of GTPs given by Bresnahan and Trajtenberg (1995) who define GPTs as pervasive, having inherent potential for technical improvements and go hand in hand with the development of innovational complementaries.

The first two indicators are mainly drawn from Teichert (2012) and adapted for application to the SKIN model. For all SKIN runs, we use the same input values for most parameters, which are (Table 9.1):

We only vary the share of big firms (either 1 or 5 %) and the partnership strategy (conservative and progressive) as these parameters have turned out to have a significant influence on the indicators investigated. The knowledge as represented as capabilities is analysed either when the simulation seems to have 'initialised' (approx. 100 ticks) or 'settled' (approx. 220 ticks), and half way in between after 160 ticks. The object of analysis is the capabilities in the model and their combination into an innovation hypothesis (IH). Furthermore, we divide the available knowledge space of 1,000 capabilities into 10 classes of 100 possible capabilities each.

9.3.2 *Pervasiveness and Capability Distribution*

To be seen as pervasive a GPT has to show a wide range of applications in various other technologies (Teichert 2012). Diffusion is one possible indicator of pervasiveness, measuring the share of innovation hypotheses carrying at least one capability of a certain class. Here, the shares of the different classes have to be compared with other technologies in order to identify whether there is a class applied significantly more often than all other classes.

Table 9.1 Standard scenario parameter settings

Number of Firms	200
Number of Products	50
Number of inputs	4
In-out-products	10
Initial capital	16,000
Attractiveness threshold	0.3
Success-threshold	1,000
Reward-to-trigger startup	1,200

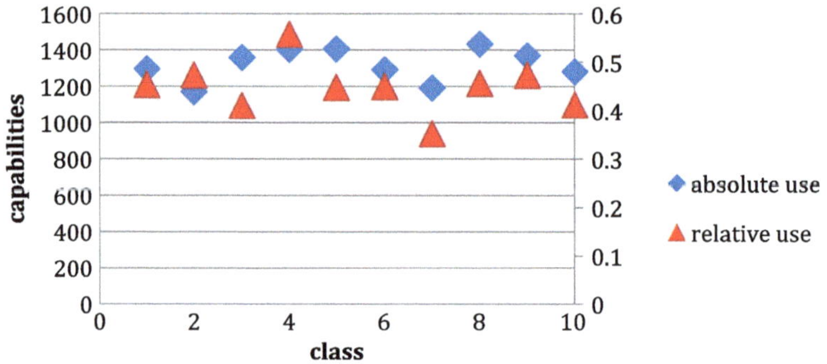

Fig. 9.3 Absolute and relative appearance of capabilities

Based on the idea of the GPT having to be combined with knowledge available in the application sector, we would assume that the GPT knowledge is not only highly present in absolute numbers but also used in the majority of IHs, or at least used by an order of magnitude more often than application technology knowledge.

In order to find out whether this is the case for certain capability classes in the SKIN model, we count for each class how often the capabilities of the respective class are used in the IHs of one simulation run at a certain time (e.g. 100 ticks). Figure 9.2 shows the absolute use (scale on the left) and diffusion (scale on the right)[1] of each capability class over all IHs after 100 ticks, averaged over 6 runs (strategy: conservative; 1 % big firms). None of the classes seems to be an out-layer, since the values appear to be equally distributed. Exemplary, Fig. 9.3 shows the absolute appearance of capabilities after 220 ticks for a progressive strategy. The absolute number of capabilities used is significantly smaller, however the pattern of a rather equal distribution is similar. Conducting the same test after 100 and 160 ticks as well as for high (5 %) and low (1 %) shares of big companies does not change this pattern.

We can therefore claim that none of the technologies is significantly utilized more or less often than any of the other technologies. As described above however, we claim that this has to be the case for a pervasive technology.

[1] Diffusion here is measured as the share of IHs containing a capability of the respective group.

The pervasiveness of a technology not only manifests itself in the diffusion rate: empirically, with a GPT being applicable in very different industries, this should also be reflected in the patents of this technology. Various authors proposed to measure this wide applicability with a generality index (Bresnahan and Trajtenberg 1995; Hall and Trajtenberg 2004; Moser and Nicholas 2004), based on the citations a patent receives from other than their own technology class. Since this measure suffers from not being a weighted measure, thus assuming all technologies being equally distant from each other (Hall and Trajtenberg 2004), we apply a weighted measure of generality exemplary for one run of the SKIN model with a standard setting after 100 ticks. This technological coherence describes the extent to which the capabilities used in the IHs are similar (see Teichert 2012). The more similar they are, the more specialised a technology class is.

First, we calculate the co-occurrence of capabilities in all innovation hypotheses in the simulation run according to Antonelli et al. (2010). The resulting matrix indicates for each capability how often the respective capability is used together with any other capability in the IHs. The technological relatedness values ($\mathrm{TR_{ij}}$) between each pair of capabilities $\mathrm{C_i}$ and $\mathrm{C_j}$ are then weighted by the relative share of IHs in the respective technologies:

$$\mathrm{TC_i} = \frac{\sum_{i \neq j} \mathrm{C_{ij}} \times \mathrm{IH_j}}{\sum_{i \neq j} \mathrm{IH_j}}$$

with $\mathrm{IH_j}$ as the number of IHs. Grouping the capabilities available in the simulation into ten classes, the $\mathrm{TC_i}$ values are added up weighing them by the relative share of IHs using the respective capability. This gives the technological coherence of each class.

Lower values signify that the capabilities in the respective class are used less often with the same capabilities from other classes in the IHs, thus they are combined with a less coherent set of other capabilities. A GPT would thus have a lower $\mathrm{TC_i}$ than a non-GPT. The absolute level of $\mathrm{TC_i}$ is therefore less important than the relation to other technologies. As shown in graph Fig. 9.4 the accumulated weighted technical coherence of the assumed ten technology classes only differ slightly (mean $\mu = 0.0303$; standard deviation $\sigma = 8.09E^{-4}$).

Again, the SKIN generates values that appear to be equally distributed, i.e. none of the technology classes is combined with a broader or narrower set of classes than all other classes (here after 220 ticks). We can conclude that none of the classes has a higher compatibility or generality than the others. The criterion of pervasiveness measured by diffusion and technological coherence is therefore not fulfilled yet in the SKIN model.

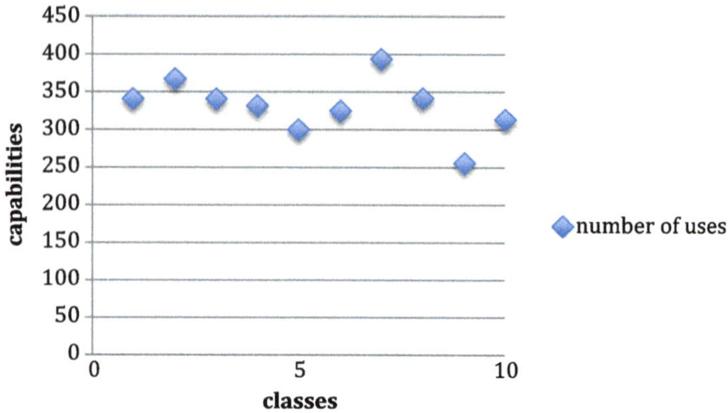

Fig. 9.4 Absolute appearance of capabilities

9.4 Merging of Knowledge and Network Structures

Analysing the SKIN model and the structure of knowledge networks it reproduces, we focus on the basic structure of knowledge implicitly described above. This means a separation of knowledge into different AS, which are only weakly connected and one or more GPT sectors. It is widely accepted that various networks of knowledge (e.g. patent co-classification networks, citation networks) show some clustering of knowledge for instance in biotech (Cooke 2002) ICT (Antonelli et al. 2010). Within these clusters or sub-networks, knowledge should be tightly connected, as the pieces of knowledge are very similar to each other and combined frequently. Between the clusters however, connections are expected to be relatively sparse. This can also be seen in the empirical data on the co-classification data from Ireland, where the knowledge of most areas clearly clusters.

One way to measure the degree to which a network is separated into densely connected sub-networks was shown by Blondel et al. (2008) who define modularity as the share of links inside the community relative to all links in the network, resulting in a scalar value between -1 and 1, with highly clustered networks having a modularity closer to 1. With densely connected application sector knowledge and only sparse connections between the AS clusters, we would expect the modularity to be close to 1.

If the knowledge in the networks is merging, for example through the appearance of a GPT, we would expect more connections between the sub-networks to become established, thus the modularity to decrease. This certainly only holds given the connections within the sub-networks do not increase at the same rate. We apply this measure to the capability network in SKIN as shown below, trying to investigate whether knowledge mergence can already be simulated in the basic SKIN model.

The knowledge network in SKIN is based on the utilisation of capabilities in IHs (Fig. 9.5). Whenever two or more capabilities are used in the same IH *(italics)*, a fully connected network between these capabilities (circles) is established:

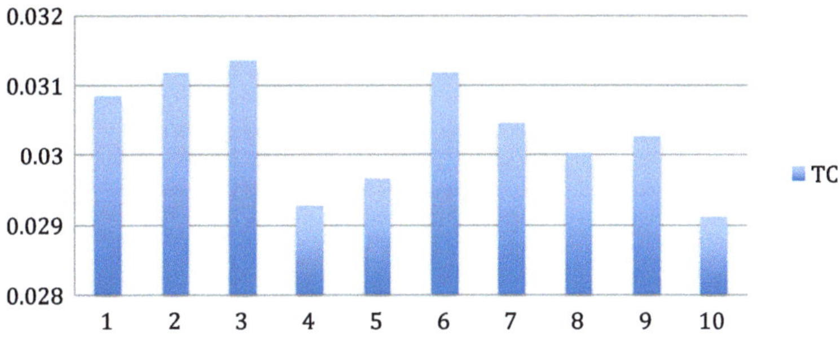

Fig. 9.5 Technological coherence of capabilities in IHs

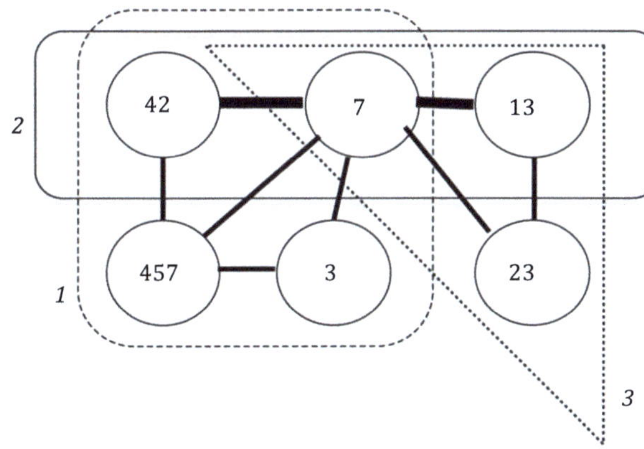

Fig. 9.6 Capability network scheme

Links between capabilities are weighted, thus the weight of the link between them will be increased whenever these capabilities are combined more than once in an IH (capabilities 42 and 7 are used in IH *1* and *2*, capabilities 7 and 13 in IH *2* and *3*).

From this, we can draw the network of capabilities for the simulation runs in SKIN as seen exemplary in Fig. 9.6. The graph shows the giant component of the capabilities network (616 nodes, 7,587 links) with conservative partnership strategy and a low share of big firms (1 %) after 160 ticks. The modularity of this network is 0.53, node and link colours correspond to the sub-network identified by the modularity algorithm.

Figure 9.7 shows a network of low modularity (progressive strategy, high number of big firms, 220 ticks) with 298 nodes and 8,958 links. It is clearly visibly that sub-networks (application sectors) are hard to detect (Fig. 9.8).

Figures in Table 9.2 show the modularity coefficient for various settings and are averaged over 5 runs per scenario. The empirical data based on co-classifications of patents with an Irish assignee gives us a modularity of 0.71, supporting our

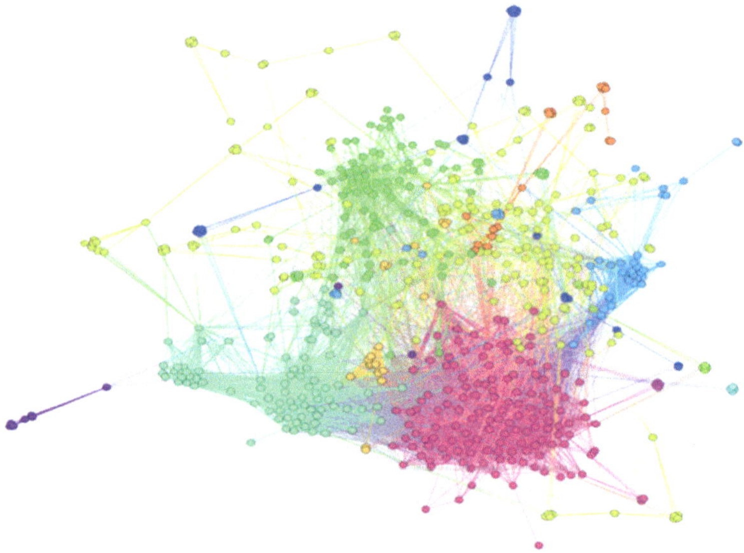

Fig. 9.7 Exemplary network of capabilities—high modularity (network graphs and analyses were conducted using Gephi)

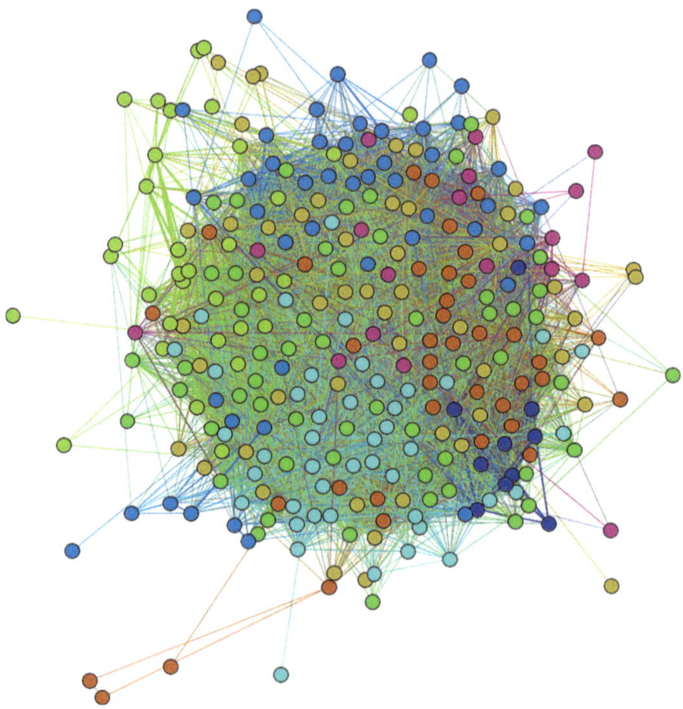

Fig. 9.8 Exemplary network of capabilities—low modularity

Table 9.2 Modularity of capability networks

	Big firms 1 %	Big firms 5 %
Empirical case	0.71	
Conservative (100/160/220)	0.524/0.534/0.479	0.589/0.498/0.288
Progressive (100/160/220)	0.181/0.197/0.184	0.23/0.174/0.155

hypothesis of a network of different pieces of knowledge, clustered in sub-networks with sparse connections.

The simulated data clearly shows the influence of the partnering strategy on the evolution of modularity. In the progressive scenarios, modularity tends to decrease faster, i.e. the knowledge merges at a higher rate. In the progressive setting companies tend to chose partners with more distant knowledge, thus connections to other sectors are established more quickly than in the conservative scenario, where companies tend to chose similar partners. This is also in line with the findings of Pyka et al. (2007), who show that firm networks in SKIN only for conservative scenarios develop scale free characteristics. Small niches of knowledge retain for a longer period since firms tend to form small world networks.

Also, the influence of the share of big firms is clearly visible. This may stem from big firms having a larger set of capabilities, thus leading to a higher probability that capabilities get combined when they are in the same firm than when they are distributed across firms.

The simulated values seem to support that knowledge merging is already emerging in the simulation. However, in the previous section the existence of GPTs in the SKIN model could not be supported. Since all characteristics of a GPT have to be fulfilled in order for a technology the be a GPT, the knowledge merging taking place cannot be traced back to the influence of GPT capabilities, but to the factors already mentioned (partnering strategy and share of big firms). Additionally, the extent to which the knowledge merges, resulting in a network of capabilities of uniform connectedness, does not seem to be supported by the empirical case.

9.4.1 Further Work: Indicators and SKIN Adaptation

9.4.1.1 Indicators

As shown with the indicators and network structure and model behaviour above, GPTs are clearly not yet emerging in the SKIN model. Changing the SKIN model in a way that it shows the behaviour described seems possible. However, further stylised facts from literature were not yet taken into account, since they are mainly based on indicators not to be implemented straight forwardly:

- GPTs show scope for on-going *technological improvement*: According to Teichert (2012), with accelerating growth rates of GPT-inventions (Palmberg and Nikulainen 2006) and forward citation rates in patents (Hall and Trajtenberg 2004) as well as in publications, at least two measures can be found in literature.

In the SKIN model, numbers and growth rates for IHs using at least one 'GPT' capability could be computed as a measure of inventions using the emerging GPT. Forward citation rates could be based on the inputs an IHs uses and the capabilities required in order to manufacture these inputs. The equally distributed use of all technologies even after a high number of ticks indicates however that none of the technology classes outperforms any other class over a longer period of time.

- GPTs *spur innovation* in application sectors: the major indicator for a GPT such as nanotechnology to spur innovations in other sectors along the value chain is the diffusion rate of the GPT knowledge (Shea et al. 2011) as well as the growth of classes citing the GPT (Hall and Trajtenberg 2004). A measure for diffusion has already been described in chapter 9.3.2. The second indicator may be compiled by identifying the growth rates in terms of IHs assigned to the classes citing or using GPT knowledge. Since this knowledge enables new innovations in the classes it is combined with, a higher growth rate of these classes as compared to other AS classes which are not (yet) combined (Teichert 2012) is to be expected.
- As stated above, one stream of GPT literature identifies the development of *complementaries* to be combined with the GPT as one major prerequisite for the GPT to unfold its economic potential. Bresnahan and Trajtenberg (1995) already introduced the concept of vertical and horizontal externalities, horizontally between different AS and vertically between the GPT sector and the various AS. The complementariness can again be computed with the help of citations relating the number of second generation citations of the GPT to all other citations not towards the GPT. In the SKIN model, input chains over two steps would have to be calculated in order to measure the complementarity index.

9.4.1.2 Knowledge and Value Chain Implementation

In order to simulate the GPT character of nanotechnology in a more sophisticated way as currently done in SKIN, the application of genetic algorithms (Birchenhall et al. 1997) seems to be a promising alternative. Products would require a combination of at least two strings of knowledge, one from the application technology and another from the GPT. Mutations of the genes could provide a source of entirely new knowledge in addition to radical research. Internal and external selection mechanisms (Geisendorf 2010) would provide a two step selection process, ensuring both, technological feasibility and marketability.

On the product level in SKIN, value chains with producers of raw materials, intermediaries and end products already exist. However, a specific GPT sector running parallel to the value chain and having links to all other steps of the value chain as described in chapter 9.2.3 is not yet implemented. This is especially important since this may have different effects in the simulation such as a direct dissemination of knowledge from the raw materials sector to the end products sector or additional sources of innovation by connecting the lateral link to the other chain links. A differentiation between GPT actors and application actors in the model would replicate the value chain more realistically and allow for different characteristics of the actors: there is evidence for example that the nano-tools sector is dominated by SMEs and start-ups, whereas in the application sector MNEs dominate the market (Gómez-Uranga et al. 2011).

9.5 Conclusions

Nanotechnology is showing strong signs of being a GPT emerging, thus generating huge interest of policy makers wanting to effectively influence nanotechnology evolution. Since there is strong evidence in literature about nanotechnology being a GPT, the uniqueness of GPTs as compared to other 'standard' technologies, makes a separate modelling attempt for nanotechnology as a GPT reasonable. Still, not every single characteristic concerning the emergence and evolution of nanotechnology can be expected being found with all other GPTs, nor vice versa.

A set of indicators to identify GPTs was developed, based on their definition. Two of the main characteristics (pervasiveness and the knowledge merging effect) were applied to the standard SKIN model. Neither the diffusion of capabilities nor their coherent/incoherent use within the model allowed for an identification of GPT knowledge in SKIN. Analysing the capability networks of SKIN not only showed signs of excessive knowledge mergence, but also this could not result from the existence of GPTs. Based on prior analyses of the SKIN model, a quicker mergence of knowledge in progressive scenarios could already be expected. The extent however, remained unclear.

Besides the indicators applied, three more stylised facts from the definition of GPTs and the respective measures are introduced. Since they all are based on citations, a meaningful representation within the SKIN model has to be identified or constructed. This would also allow us to make use of more network indicators such as in-out degree distributions and clustering coefficients. For the purpose of modelling two distinct types of knowledge with different characteristics, the introduction of a more sophisticated knowledge representation, combination and selection is proposed with the genetic algorithms. Finally, improvements for the value chain are proposed.

The SKIN model may not yet be a meaningful simulation when it comes to general purpose technologies such as nanotechnology, however it provides useful starting ground for the simulation of GPTs from a knowledge-based perspective for which we have introduced the indicators necessary to identify GPTs in SKIN.

References

Aghion P, Howitt P (1998) On the macroeconomic effects of major technological change. In: Helpman E (ed) General purpose technologies and economic growth. MIT Press, Cambridge, MA

Antonelli C, Krafft J, Quatraro F (2010) Recombinant knowledge and growth: the case of ICTs. Struct Change Econ Dynam 21(1):50–69

Bhattacharya S, Shilpa M (2011) Mapping nanotechnology research and innovation in India. DESIDOC J Libr Inform Technol 31(5):349–358

Birchenhall C, Kastrinos N, Metcalfe JS (1997) Genetic algorithms in evolutionary modelling. J Evol Econ 7(4):375–393

Blondel VD, Guillaume JL, Lambiotte R, Lefebvre E (2008) Fast unfolding of communities in large networks. J Stat Mech Theory Exp 2008(10): P10008. doi:10.1088/1742-5468/2008/ 10/p10008

Bresnahan TF (2010) General purpose technologies. In: Hall BH, Rosenberg N (eds) Handbook of the economics of innovation. Elsevier-North-Holland, Amsterdam

Bresnahan TF, Trajtenberg M (1995) General purpose technologies "engines of growth"? J Econ 65(1):83–108

Carlaw KI, Lipsey RG (2006) GPT-driven, endogenous growth. Econ J 116:155–174

Cooke P (2002) Biotechnology clusters as regional, sectoral innovation systems. Int Reg Sci Rev 25(1):8–37

Forfás (2010) Ireland's nanotechnology commercialisation framework 2010–2014. Venture capital

Geisendorf S (2010) Internal selection and market selection in economic genetic algorithms. J Evol Econ 21(5):817–841

Genet C, Errabi K, Gauthier C (2012) Which model of technology transfer for nanotechnology? A comparison with biotech and microelectronics. Technovation 32(3–4):205–215

Gómez-Uranga M, Etxebarria G, Barrutia J (2011) The dynamics of regional clusters of nanotechnologies: evidences from German länder and two Spanish autonomous regions

Hall BH, Trajtenberg M (2004) Uncovering GPTs with patent data

Helpman E, Trajtenberg M (1994) A time to sow and a time to reap: growth based on general purpose technologies

Lipsey RG, Bekar C, Carlaw KI (1998) What requires explanation. In: Helpman E (ed) General purpose technologies and economic growth. MIT Press, Cambridge, MA

Lipsey RG, Carlaw KI, Bekar C (2005) Economic transformations: general purpose technologies, and long-term economic growth. Oxford University Press, Oxford

Moser P, Nicholas T (2004) Was electricity a general purpose technology? Evidence from historical patent citations. Am Econ Rev 94(2):388–394

Nikulainen T (2010) Identifying nanotechnological linkages in the Finnish economy–An explorative study. Tech Anal Strat Manag 22(5):513–531

OECD (2009) OECD science, technology and industry scoreboard 2009. OECD, Paris

Palmberg C, Nikulainen T (2006) Nanotechnology as a general purpose technology of the 21st century? An overview with focus on Finland. DIME working paper series

Porter AL, Youtie J (2009) How interdisciplinary is nanotechnology? J Nanopart Res 11(5): 1023–1041

Pyka A, Gilbert N, Ahrweiler P (2007) Simulating knowledge-generation and distribution processes in innovation collaborations and networks. Cybern Syst 38(7):667–693

Schiess D (2011) Invention and diffusion of general purpose technologies. ETH Zurich

Shea CM, Grinde R, Elmslie B (2011) Nanotechnology as general-purpose technology: empirical evidence and implications. Tech Anal Strat Manag 23(2):175–192

Solow R (1987) We'd better watch out. New York Times Book Review, p 36

Teichert N (2012) Innovation in general purpose technologies: how knowledge gains when it is shared. Karlsruher Institut für Technologie (KIT)

Van Zon A, Fortune E, Kronenberg T (2003) How to sow and reap as you go: a simple model of cyclical endogenous growth. Maastricht Economic Research Institute on Innovation and Technology, Maastricht

Youtie J, Iacopetta M, Graham S (2007) Assessing the nature of nanotechnology: can we uncover an emerging general purpose technology? J Technol Transf 33(3):315–329

Chapter 10
Multilevel Analysis of Industrial Clusters: Actors, Intentions and Randomness Model

Ozge Dilaver, Elvira Uyarra, and Mercedes Bleda

> *When an industry has thus chosen a locality for itself, it is likely to stay there long: so great are the advantages which people following the same skilled trade get from near neighbourhood to one another. The mysteries of the trade become no mysteries; but are as it were in the air, and children learn many of them unconsciously. Good work is rightly appreciated, inventions and improvements in machinery, in processes and the general organization of the business have their merits promptly discussed: if one man starts a new idea, it is taken up by others and combined with suggestions of their own; and thus it becomes the source of further new ideas. And presently subsidiary trades grow up in the neighbourhood, supplying it with implements and materials, organizing its traffic, and in many ways conducing to the economy of its material (Marshall 1920, Book IV, p 27).*

Abstract The literature on industrial clusters indicates a symbiotic relationship between innovation and geographical concentration of firms working in similar industries. Innovative processes require different forms of knowledge and expertise, which are distributed across individuals and organisations at different levels of industrial clusters. In this chapter, we present fundamental extensions to the SKIN model for representing such multi-level interactions. We introduce individual actors in addition to firms as agents. These agents are placed in a two-regions environment that simulates evolution of two competing regions. We also integrate elements of intentionality in addition to randomness in our model. Through subjective assessments of their managers, firms investigate and design research

---

O. Dilaver (✉)
Centre for Research in Social Simulation (CRESS), University of Surrey, Guildford, Surrey GU2 7XH, UK

British Institute at Ankara, Tahran Caddesi 24, 06700 Ankara, Turkey
e-mail: o.dilaverkalkan@surrey.ac.uk

E. Uyarra • M. Bleda
Manchester Institute of Innovation Research (MIoIR), Manchester Business School, University of Manchester, Oxford Road, Manchester M13 9PL, UK
e-mail: Elvira.Uyarra@mbs.ac.uk; Mercedes.Bleda@mbs.ac.uk

N. Gilbert et al. (eds.), *Simulating Knowledge Dynamics in Innovation Networks*, Understanding Complex Systems, DOI 10.1007/978-3-662-43508-3_10,

projects. These extensions help to open up the black box of the firm and relate firms to the creative agency of individuals in starting up new firms, establishing their research objectives and creating new knowledge. Within this broad range of issues we focus in this chapter on the role of entrepreneurship to illustrate how the extended model can be used. In experiments focusing on entrepreneurship, we generate the relative success of Silicon Valley in comparison to Boston *in silico*.

10.1 Introduction

Innovative processes require different forms of knowledge and expertise, both of which are distributed across individuals and organisations at different levels of a socio-economic system. Collaborative interactions between these agents not only enable the sharing of knowledge, expertise and workload but also facilitate the identification of new areas where existing technologies and new forms of knowledge can be used. These interactions do not occur in a vacuum; they have *socio-temporal* and *spatial* properties.

Collaborative interactions have *socio-temporal* properties because they are continuously shaped and re-defined by social contexts and common habits of thought. In addition to this shared background, interactions between two parties may have their own history. When relatively stable patterns of interactions are made sense of through shared understandings of social institutions, roles and expectations, they are known as relationships. At the macro level, the occurrence of successful collaborations over time gradually form distinctive interactive structures known as innovation networks (Breschi and Malerba 2005; Pyka 2007), consisting of inter-personal and inter-organisational relationships and shared habits of thought that underpin these relationships. As such, the innovative behaviour of individuals and organizations are path-dependent and socially embedded.

Regarding the *spatial properties* of innovative interactions, recent literature has highlighted the role of geography in shaping knowledge interactions (Glückler 2007). Given the aforementioned socially embedded and continuously evolving nature of innovation networks, it is not surprising that geographical proximity constitutes a facilitating factor in knowledge exchanges. Scholarly research on industrial districts, (Pyke et al. 1990), clusters and regional systems of innovation (Cooke et al. 1997) reflect a symbiotic relationship between innovation, networks, and geographical proximity in regional development. Regions displaying dense networks of inter-firm and public-private interactions are associated with better than average growth and innovation performance in this literature (Cooke and Morgan 1993). As a result, understanding the conditions and the processes underpinning the emergence of these networks or clusters has captured the interest of academics and policy makers alike.

Indeed, policy makers have repeatedly sought to replicate network configurations identified in successful regions. Yet, and despite the academic and policy interest placed on the key influence of clusters, there is still a limited understanding of how they "emerge, take hold and transform regional economies" (Feldman et al. 2005; p. 130). The emergence and evolution of regional innovation networks are shaped by the nature and dynamics of the complex organizational and institutional interrelations they enable. The intrinsic complexity of these networks makes it difficult for traditional mathematical and statistical modelling techniques to provide results and robust explanations of their dynamics.

Computer simulations have, on the other hand, been successfully used to analyse the emergence of complex innovation networks, and have been able to reproduce a number of stylised facts on their dynamics which have not been well accounted for by conventional mathematical formal tools. The Simulating Knowledge Dynamics in Innovation Networks (SKIN) model (Ahrweiler et al. 2004) is one of the well-established agent-based models, which, as exemplified in various applications in this volume, has been used for investigating the key factors shaping the formation of innovation networks. Innovation networks in the SKIN model are formed by firms that possess different knowledge bases, and evolve over time via interactions with the environment. Thus, SKIN enables the study of the emergence, adaptation and co-evolution of innovation networks and the knowledge bases of their participant firms.

In this chapter, our aim is to contribute to a better understanding of the dynamics of industrial clusters. In particular, we aim to understand why in the absence of clear natural advantages, some regions establish denser and more successful innovation networks than others. We present a fundamental extension to the SKIN model that we consider is highly relevant to address our research question. Our extended model employs a multi-level modelling approach where we represent both firms and actors (individuals) as agents. This extension provides a micro-level basis to our analysis, enables us to study the interactions between actors within and outside the firm and give us an angle to integrate elements of intentionality in the model, which otherwise relied on stochastic processes to represent innovative processes. In particular, we focus on the role of entrepreneurs are key innovative agents and critical elements in the formation and viability of innovate networks and clusters (Feldman et al. 2005).

The chapter is organised as follows. In the next section, drawing upon the existing literature, we present key constructs and findings about industrial clusters and regional innovation networks. The third section presents a detailed description of our multilevel modelling approach, introduces the model and illustrates potential uses of the model with experiments studying the role of entrepreneurship in the evolution of networks in two competing hypothetical regions. The fourth section concludes.

10.2 Firms, Innovation Networks and Industrial Clusters

10.2.1 Innovation as Processes of Collective Understanding

As we have indicated in the introduction, innovation generally occurs in the context of both socially and spatially embedded networks. While established networks shape innovation processes in fundamental ways, it is the individuals and the organizations forming these networks that ultimately constitute the origin of any processes of novelty and knowledge creation. One important aspect of this knowledge creation is the collective efforts of dealing with uncertainties involved in innovative processes. It is widely accepted in innovation studies that innovative processes are inherently uncertain and innovators and entrepreneurs attempting to develop and supply innovations face uncertainties related both to technical and commercial success of their activities (Schumpeter 1939; Mises 1998 [1949]; Rosenberg 2003; Jalonen 2012).

Innovations can lead to changes in beliefs, habits and routines used by individuals and organizations to retain elements of consistency and stability in their environments. Lane and Maxfield (2005) study the uncertainty that characterise the cognitive and social construction processes of knowledge creation associated with the generation of innovation. The authors argue, in particular referring to case studies from Silicon Valley, that innovative processes often involve ontological uncertainty that agents can only cope with by using a *narrative logic*. Accordingly, agents hold uncertainty temporarily at bay by interpreting their contexts and their own roles in relation to familiar stories based on their past. Although such narratives are subjective they work on a background of collective understandings. Agents continuously interpret developments in the light of their narrative and can act as if they know more than they actually do. Indeed, innovative processes often take place in 'thought environments', since innovations, by their very nature, bring about significant levels of novelty and newly created knowledge, which remain to be experienced and made sense of.

Uncertainty also characterises innovative processes at the organizational level, where learning strategies and procedures of searching for new solutions have to cope with ignorance and fundamental uncertainty. The interdisciplinary dichotomy between *exploration* and *exploitation* (Schumpeter 1934; Holland 1975; March 1991) has often been referred to in this context. The dichotomy separates highly uncertain endeavours for acquiring knowledge and awareness about the space of possibilities, from laborious implementations, corrections and refinements of solutions that were already known, albeit more crudely. The need for finding a strategic balance between explorative and exploitative activities is emphasised both in innovation and organisational learning literatures. It is argued that too much exploration may lead to many underdeveloped ideas with lack of distinctive competencies whereas too much uncertainty and exploitation may trap agents in suboptimal solutions (March 1991), leaving them vulnerable to the action of other innovators.

10.2.2 Firms and Innovation Networks

The uncertainty involved in knowledge creation is also highlighted in complex evolutionary accounts of the firm. These accounts approach firms as networks of multiple individuals (interlinked within specific organisational structures) that interact within an environment, converting information into useful knowledge in order to increase firm performance (see e.g. Foster 2005). In this perspective, innovative processes are considered highly uncertain partly because in a complex evolutionary setting, knowledge is not fully codified, but partly tacit and firm specific. Knowledge is also limited because firms are boundedly rational (Cyert and March 1963; Simon 1976), that is: (1) they usually do not hold a perfect or complete understanding, or accurate models of the reality or environment they are embedded in, and (2) they experience limits in the processing of information from the environment as well as in formulation and solving of complex problems.

As a result, the creation of new knowledge via organizational learning is context-specific, path dependent and highly idiosyncratic (Hodgson 1998). Hence, two firms endowed with the same physical assets and informational resources operating in similar environments will hold different sets of knowledge (including their abilities and understandings of the environment) and most likely do things in different ways—(Malerba and Orsenigo 2000). A fundamental implication of this is that knowledge and the capability to create and utilise such knowledge constitute the primary sources of both heterogeneity and differential competitiveness among firms (Prahalad and Hamel 1990; Nelson 1991; Teece et al. 1997). Differences in firms' performance rest to a great extent on idiosyncratic features that are related to the creation of their specific knowledge as well as the path dependent process by which this knowledge is modified and built up over time through learning (Nelson 1991; Hodgson 1998; Nonaka et al. 2000).

Differences in firm's innovative behaviour and performance have also their source in the structure of interactions among the individual constituents of firms. Individuals in a firm are interlinked in a connective organisational structure within the firm, and apply their skills to generate behaviour through their mutual interaction (Nonaka et al. 2000). Thus, their contribution to the firm's collective behaviour and performance is shaped and dependent on the structure of interactions (Dosi and Marengo 1994; Marengo 1996), which is intrinsically linked to the formal organisational structure of the firm (Marengo 1992). While a firm's agency is often thought through decisions of top management level, within the network structure of a firm, order and firm-level behaviour can emerge from the bottom up from the local interactions at the micro level of its components. In this context, the highly specific nature of the configuration of interactions among the members of a firm determine to a great extent the collective behaviour of the firm. Such multilevel perspectives of the firm, therefore, can provide solid micro foundations for explanations of the collective innovative behaviour of the firm.

10.2.3 Innovation Networks and Space

As mentioned in the introduction, innovation networks constitute an additional fundamental factor shaping and facilitating innovation processes in firms. They are not only socially and institutionally embedded but they also have geographical and spatial properties that are influential in their formation and long-term dynamics. This localised nature of innovation networks has long captured the interest of scholars in economic geography and innovation studies. 'Knowledge spillovers' and related concepts suggest a 'typical distance decay function in communication' (Howells 1999) due to the tacit nature of knowledge.

Economic geographers studying industrial agglomerations such as so-called 'industrial districts' in Italy have stressed the positive independencies between innovation, networks, and proximity, by highlighting the localized nature of inter-firm networks of cooperation and their influence on regional competitiveness (Koschatzky and Sternberg 2000). Discussions on inter-firm networks in economic geography have generally focused on the institutional context in which actors are embedded, rather than on the networks and actors per se. Such explanations generally see institutions as exogenous rather than as evolving, adaptive social constructs and tend to downplay the agency of individual and collective actors, entrepreneurs in particular (Feldman et al. 2005).

Existing studies also tend to neglect multi-level analyses that incorporate both entrepreneurs and firms (Sternberg 2009). Ideas of clusters and industrial districts have generally stressed the benefits of being located in a cluster and treated clusters as the result of location choices. For instance, according to Krugman (1991), and given the external economies which arise from co-location between firms in complementary activities (such as better access to specialised suppliers, labour pooling and knowledge spillovers) firms would choose to locate close to other firms in order to exploit the benefits of clusters. However it is dubious whether firms actually engage in such conscious searching and comparing activities (Zhang 2003), locational choices being more likely to be driven by entrepreneurs' choices and dominated by inertia.

The importance of entrepreneurs as change agents justifies their inclusion as units of analysis together with firms in explaining innovative agglomerations. For instance Audretsch (1995) highlights the role of individuals endowed with new economic knowledge, such as scientists and engineers, who may decide to pursue a new idea outside of the organizational structure of the firm and start a new firm to appropriate the value of this new knowledge.

Empirical studies on entrepreneurial activity have increasingly argued that entrepreneurship is a "regional event" (Feldman 2001) given the localized nature of the determinants, processes and impacts of entrepreneurial actions (for a review see Sternberg 2009). Regional determinants of the number of start-ups in a region include availability of factors such as skilled labour (Doms et al. 2010) and local universities (see e.g. Anselin et al. 1997), the presence of certain support institutions and policy (such as the provision of venture capital) and inclinations for certain organisational structures.

Entrepreneurship is also a local event because entrepreneurs tend to start new ventures at places where they previously lived and worked (Stam 2010; Audretsch et al. 2012). Locational inertia occurs because entrepreneurs can benefit from their existing networks and access to resources more easily at their home locations (Feldman et al. 2005; Stam 2010). As Schumpeter (1934) argued, "success breeds success, and the appearance of one or a few entrepreneurs facilitates the appearance of others" (p. 228). Entrepreneurs may be following signs of success and early stories of successful entrepreneurs may encourage others to start-up businesses in similar or complementary activities. Holcombe (2003) similarly argues that entrepreneurship enlarges the stock of entrepreneurial opportunities by creating new market possibilities.

For Sorenson and Audia (2000), changes occurring in the structure of entrepreneurial opportunities lead to increased founding rates. Industrial clustering therefore occurs largely via above average firm founding rates sustained by significantly high levels of entrepreneurial activity. At the same time, Sorenson and Audia (2000) argue that structurally equivalent organizations also compete with each other for vital resources. Co-location increases competition for the same pool of human resources or product markets and therefore leads to higher rates of failure among competing organisations. Looking at the shoe industry, they conclude that it is variation in the structure of entrepreneurial opportunities that maintains geographic concentration in the sector.

Culture has also been put forward as potential explanation for the spatial variation of clusters. This includes a culture of entrepreneurship (Hofstede 2001) acceptance as opposed to the 'stigma' of failure (Landier 2001), tolerance and creativity (Florida 2005) of certain locations. These contextual factors are often summed up as 'entrepreneurial climate'. As noted by Stam (2010; p. 149), this institutional environment acts as a "mediating factor between investments in the knowledge base of a society and the knowledge spill-overs exploited by entrepreneurs." It has been argued however, that institutions should not be treated as exogenous but as evolving, adaptive social constructs (Feldman et al. 2005, see also Boschma and Frenken 2009). Feldman et al. (2005; p. 130) and Stam (2010) describe how initial concentrations of economic activity are often accompanied by a (conscious or unintended) formation of dedicated supporting institutions, in a co-evolutionary process of technology, market and institutional development. Subsequent growth of a cluster would in turn be reinforced by the presence of such supporting institutions (Maskell and Malmberg 2007).

Hence, entrepreneurship is the result of the interaction between individual attributes and the surrounding environment where opportunities are likely to arise (Stam 2010). The surrounding environment shapes but is also shaped by the actions of entrepreneurs. Entrepreneurship is incumbent upon the presence of opportunities and by individuals making decisions and acting upon such technological and market opportunities.

10.2.4 Simulation Models of Innovative Processes

Given the multi-level and complex nature of all the above factors and their influence in the dynamic formation and evolution of innovation networks and clusters, traditional mathematical and statistical modelling techniques are generally not appropriate for their analysis. As indicated in the introduction, agent-based models constitute a useful tool to address the complexity of innovation networks and industrial clusters as they are intrinsically multi-level and hence appropriate for an investigation of the macro-level outcomes of micro-level interactions. Clusters and networks have been approached from a number of different angles in simulation studies.

One stream of simulation studies has investigated the evolution of inter-firm networks by analysing the firms' decisions to establish connections with other firms. These studies suggest that a firm may be particularly attractive for some firms and not to others depending on the match between their knowledge bases, their previous relationships and their relative positions across business networks. These conditions, known as cognitive, relational and structural embeddedness between firms, break the anonymity of firms in alliance selection and accommodate the formation of path-dependent structures. Different combinations of such factors can lead to temporal or industry-related idiosyncrasies in network structures (Cowan et al. 2005). Another strand of models has investigated how different network structures affect innovative activities and diffusion of knowledge. Cowan and Jonard (2004), for example, study diffusion of knowledge in social networks that have small-world (Watts and Strogatz 1998) properties.

Regarding spatial patterns and agglomeration, a number of authors have developed simulation models. Brenner (2001), for instance, has built a model for studying localized dynamics leading to the development of industrial clusters, and found that a broad range of factors including start up rate, radical innovations, availability of experience employees, number of spin-offs, venture capital and technological synergies can explain cluster emergence. Similarly, in Otter et al. (2001) emergence of clusters is sensitive to initial conditions related to factor endowments as well as exogenous conditions. Boero et al. (2004) developed an agent-based prototype that has been created to investigate the relation among local labour markets, entrepreneurship and human capital in industrial districts.

There have also been history-friendly studies that aim to re-generate existing clusters 'in-silico'. Fioretti (2001), for example developed a spatial agent-based model to study the formation and the evolution of the Prato textile district. The model reproduces the structure of interactions among firms during a 40-year period and highlights a structural transformation in the district from price based competition to one based on product variety. Similarly, Zhang (2003) built an agent-based model to study the formation of high-tech industrial clusters such as Silicon Valley. He argues that rather than moving to industrial clusters, increased entrepreneurship is the result of contagion effects, leading to greater firm creation.

Finally, the SKIN model (Ahrweiler et al. 2004) on which our work is based, analyse the evolution of innovation networks by representing firms with distinct knowledge bases (depending on the capability, ability and expertise level of the firm). In SKIN firms' knowledge bases evolve through mutations and crossovers. The model enables the investigation of collaborative interactions among firms and how these interactions can produce distinct network structures over time under different experiment scenarios.

10.3 Modelling the Emergence of Industrial Clusters

In this chapter, we describe fundamental extensions to the SKIN model (Ahrweiler et al. 2004) and this section presents these extensions. Our aim is to develop in stages a modular platform that reflects findings in different fields. We also aim to introduce actors in addition to firms as agents and integrate elements of agent's intentionality into the model while still allowing room for randomness.

We, therefore, called our extended model Actors, Intentions and Randomness, or AIR, in short. This abbreviation follows nicely the Marshallian (1920) idea that *there is something in the air* in regions that accommodate industrial clusters. In the quote we started this chapter with Marshall mentions a broad range of issues from availability, appreciation and learning of skills to effects of novelties and supply chain relationships. With our extended model, we aim to address this complexity and reveal *what it is in the air* of these regions.

In this section we first explain our general vision and directions pointing out how they relate to the key concepts and constructs in the existing innovation literature. The section then introduces our extension model AIR in detail. The third and last subsection illustrates the potential uses of AIR with experiments on two competing hypothetical regions.

10.3.1 Extending the Conceptual Model

Our extensions to the SKIN model are relatively more comprehensive than usual extensions in the modelling literature. We aim to keep well-tested strengths of the original model, while envisaging a much broader and flexible conceptual framework that can accommodate interdisciplinary depth and richness of innovation studies. We, therefore, explain modelling ideas we follow and novelties we introduce to the SKIN model at a general level before introducing AIR.

Modular Modelling An old modelling debate that is familiar to most agent-based modellers is whether the models should be simple or descriptive. While simple models help clarity, particularly in relation to implications of experiment results, descriptive models enable a more direct representation of reality and shift the research focus from abstract to relatively more concrete questions. In terms of communicating models to other academics and hence testing theoretical and constructive validity of the model, descriptive and more complicated models are difficult to disseminate. Simple models have advantages in terms of dissemination, as they have a shorter list of assumptions to remember and they can fit the journal article format more easily. However, as a stream of studies develop and numerous simple models are built on related subjects, it is not easy to derive overall contributions of these models as their assumptions and conceptual frameworks are not necessarily compatible.

We aim to develop a detailed simulation model that can address the complexity of issues related to industrial clusters and reflect knowledge accumulated in different research fields. Our aim in developing such a detailed model is being able to combine assumptions and insights of these fields and experiment their cumulative effects. On the other hand, we expect that a model that can address the multifaceted nature and complexity of industrial clusters will be complicated and require extensive research effort. We also aim to be able to communicate our model in parts, so that the model can easily be understood, critiqued and improved by other researchers.

We, therefore, aim to develop a flexible modelling structure that enables gradual development of the model through multiple research projects. We envisage handling the modelling requirements of such a comprehensive model by putting together relatively simple agent-based models that are developed for more specific research objectives. Our modelling approach, therefore, is modular.

It is common in agent-based modelling to build a simple model first and to add additional features in a stepwise manner, verifying at each step that the new versions of the model works consistently. Once completed, many features of the model can be switched on and off in order to run comparative experiments. In these respects, it can be argued that agent-based modelling is inherently modular.

Our approach complements this property of agent-based models. The element of novelty here is that we first envisage a core framework that represent some of the most important elements of environments, agents and events related to innovative processes in regional settings and that can accommodate further development of particular aspects of the model. We, therefore, first envisage an empty shelf constructed based on our reflections on the relevant literature. We then concentrate on more specific elements of the model, reflect state-of-art understanding in these particular matters, build relatively simple models and place them in our framework, like putting carefully selected books on the shelf.

More specifically, in light of the literature review presented above, we conceptualise the environment of agents, firms and actors alike, to consist of (1) uncertain rewards to their innovative activities, (2) dynamics of knowledge that different agents hold and (3) emerging social structures and constructions in regions. We represent these elements of the agents' environment with three spaces in the model, namely the reward space, the knowledge space and the earth space that we will introduce in detail in the following sub-section. Our aim is focusing on developing these spaces further in stages while tackling specific research questions related to emergence of industrial clusters.

Multilevel Structure The agents in the original SKIN model are firms. As indicated, we employ a multi-level modelling approach incorporating not only firms as unit of analysis but also representing individual actors and their interactions within and outside the firm. Actors are managers, researchers and engineers working on innovative projects, they, therefore, represent the human side of innovative processes. In this respect, we are adopting the same view as the complex evolutionary perspective outlined in Sect. 10.2. This additional level of agency enriches the analysis by opening the black box of firms and allowing the representation of the innovative processes within firms. A multi-level understanding of these innovative processes is central in our approach because the knowledge created during these processes exists in a distributed manner in the firm's codes, documents and procedures as well as in the minds of its employees.

In addition, our multi-level approach enables representing, albeit simplistically, the organisational structure of the firm, key determinant of its strategic behaviour and long term performance in an evolutionary context, and thereby truly distinguishing between networks and hierarchies at the regional level. By distinguishing between firms and actors (individuals), we can follow the impact of unsuccessful projects and firms in the knowledge accumulation in the region. Whereas in single-level models of innovation networks, unsuccessful firms and their employees disappear from the model, in our model employees continue working for other firms keeping some of the knowledge created. Our aim is to get a grasp of these multifaceted connections in our model, bridging different research streams in organisational learning, evolutionary economic geography and computational social science.

Our third purpose in incorporating the additional level of agency is modelling researchers and engineers as essential inputs of the innovative processes. Although firms can codify some knowledge and experiences in documents, work environment and procedures, most knowledge outputs of innovative activities remain in the minds and developing expertise of their employees. In our analysis it is the members of the firm, and more specifically its researchers, who connect existing knowledge and experiences and produce new ones. We think of researchers as communicating knowledge in their interactions with their colleagues and other social contacts. These exchanges have an important role in knowledge accumulation within firms and regions. Our multi-level perspective will also enable us to tackle the connections between labour market institutions that shape the work lives of researchers and the development of the innovative potential of regions. The importance of this issue is not unknown but until recently the role of labour market institutions has not been adequately addressed in simulation models of innovation.

Overall, we represent actors as the collective agency behind the firms, the major agency of innovative activities and as units or nodes of knowledge transmission within networks. Our perspective allows tackling different management and organisational structures and learning at the firm level and different intra-organisational institutions including the labour market on the development of regional systems of innovation.

Intentionality Another major extension we introduce to the conceptual framework of SKIN is modelling intentionality of agents together with the perceived randomness of social reality. The SKIN model follows an evolutionary approach, where a high number of instances with different attributes and behavioural procedures are created and allowed to interact. During the simulation experiments, some instances are left out of the simulation while new ones are entering. Over time, the population distribution of instances evolves together with structures and memories they create.

The general idea in this approach is to observe the evolution and emergence of such changes, starting from relatively simple, random or symmetric initial conditions. Models employing this line of thinking often demonstrate what processes consisting of simple interactions are capable of, *even when* agents do not directly aim for that particular outcome. It is, therefore, quite common that agents in these models start their simulated lives with random search and match processes (if any) and continue with simplistic rules, possibly dependent on their behaviour and memory in the previous periods.

In our model, we attempt to depart from this particular way of experimentation and move towards modelling innovation-related decision processes of agents more realistically. We aim to address the connections between inherently uncertain nature of innovation activities with the actors' (albeit limited) abilities to gather information, build expectations and plan research and development projects. Hence, we envisage a semi-visible innovation environment where some imperfect and subjective understandings of possible rewards from innovative activities can be built. Agents can establish targets or projects aiming specifically towards those rewards.

That being said, we are not at all aiming for a deterministic model, where agents operate through optimisation. As they are bounded rational, actors can make mistakes in understanding the rewards to or requirements of innovations and the firm can produce unplanned inventions. There are, therefore, probabilistic rather than deterministic relationships between the intentions and attributes of the actors and the outcomes of their firms' innovative activities.

10.3.2 AIR Model

In the previous subsection, we defined our modelling ideas and perspectives for extending the SKIN model. In this subsection, we describe specific properties of the core structure of AIR and we clarify its main assumptions. The simulation experiments in AIR start with the creation of the agents and the simulation environment. We will start introducing the model by explaining these elements as well.

As indicated in the previous subsection, AIR develops a multi-level structure. The classes of agents in the model are defined at three levels: actors (individuals), firms and regions. That being said, regions in the AIR can be thought both as part of the agency structure and as part of the environment where firms and actors operate. As agents, regions can be initialised with different attributes and develop different structures, networks and attributes related to innovation during simulations. As part of the environment, regions contain the firms and actors that a particular agent is likely to interact with in the short run.

Each actor in the AIR model has a knowledge base, or a set of kenes like the kenes that firms have in the SKIN model. One component of each actor's knowledge base is his or her main expertise while other components represent additional knowledge in other areas. Actors have two roles in the model: they can be managers or researchers. Managers' key function is exploring possible research projects while researchers work in research projects and construct knowledge. When working on a research project, actors' kenes develop and they may acquire bits of knowledge from each other as well. These interpersonal knowledge exchanges occur through networks of both professional and personal relationships.

Firms in the AIR model are conceptualised as entities that employ actors who actively seek for and assess innovation opportunities and run projects for realising these potentials. Firms need to complete successful innovation projects to obtain funds and pay for the salary of their employees. Firms that use all their capital go bankrupt, and their employees look for employment elsewhere, bringing their improved kenes to their new workplaces.

We conceptualise the environment of agents to consist of (1) emerging social structures and constructions in regions (2) dynamics of knowledge that different agents hold and (3) uncertain rewards to innovative activities. We represent these elements of the agents' environment with three spaces in the model, namely the earth space, the knowledge space and the reward space (Fig. 10.1).

All three spaces are envisaged in three dimensions. While the x and y dimensions of the earth space are constructed to facilitate spatial representations of agents' environment, the reward and knowledge spaces remain more abstract. The x and y coordinates of both the reward and knowledge spaces correspond to the scientific field and the application area of a project, innovation or unit of knowledge. The z coordinate of the reward space shows the magnitude of monetary rewards obtained from successful innovations and the same coordinate for the knowledge space shows the magnitude of knowledge.

In the earth space, we aim to represent initial conditions, structures and institutions in a region, and follow the evolution of these regional elements as agents produce knowledge and achieve returns over time. The earth space shows the existing firms in a region, roughly sketches the collaborations between firms, and facilitates visualisations of the evolution of industrial clusters. The earth space represents an abstract geography with two regions. This two-regions structure facilitates comparative experiments (Fig. 10.2).

In the knowledge space, we address types and properties of knowledge, and explore the creation of knowledge within and between firms. We envisage the knowledge bases of firms to be a cumulative product of the knowledge and expertise of its current and past employees. In this setting the actors generate new knowledge during their employment at research projects in collaboration with their colleagues and with the influence of their other contacts.

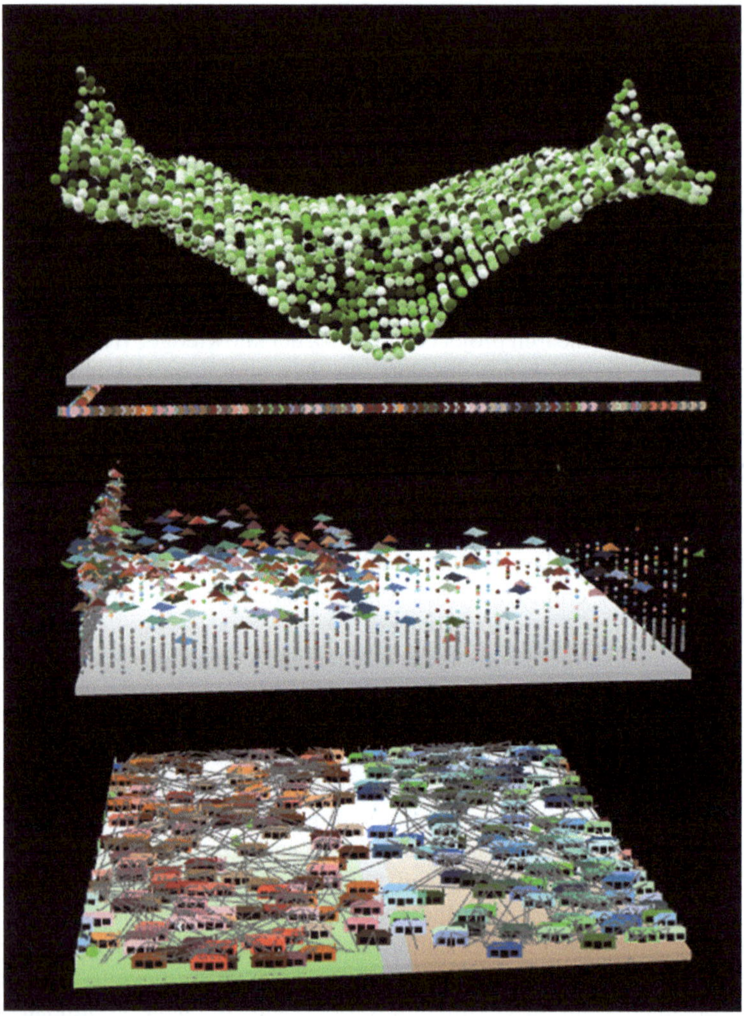

Fig. 10.1 Earth space with two regions, firms and researchers

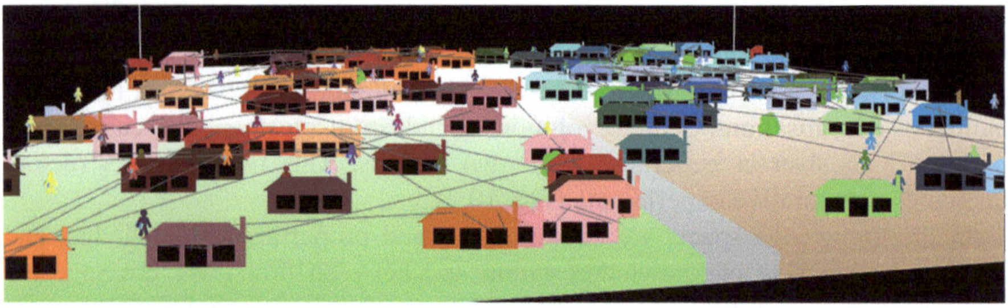

Fig. 10.2 Knowledge space with expertises of researchers (*circles*) and knowledge base of firms (*arrow heads*)

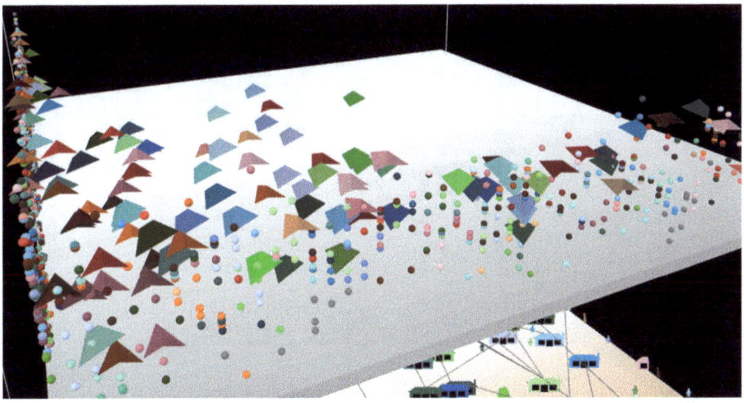

Fig. 10.3 The reward, knowledge and earth spaces

Actors are assumed to have a main expertise either in a research field or an application area and so, their knowledge bases are shown along the x and y axes of the knowledge space. Firms' knowledge bases are shown between indicated knowledge bases of their employees. If employees are from two specialties, the firm's knowledge base is shown in the midpoint between these two specialties. If they possess more than two specialties, the firm's knowledge base is shown at the centroid of the minimum simple convex polygon area that covers points representing all these specialties (Fig. 10.3).

In the reward space, we tackle the inherently uncertain nature of the returns to innovation, and the subjective understandings of these returns. We construct the reward space as a semi-visible fitness landscape that agents can collect imperfect information about. Thus, agents are bounded rational: their understanding of the reward space is not only imperfect but also subjective as their knowledge in a particular subject influences how well they can predict the returns to relevant innovations.

More specifically, the reward space contains passive but dynamic entities representing the rewards for innovation. The rewards as simulation objects hold information on a number of knowledge creation requirements. These conditions specify the magnitude of knowledge required in the main research field and the application area (x and y coordinates of the reward) and a few, relatively minor knowledge creation requirements in other fields and areas. The fulfilment of all such knowledge creation requirements corresponds to an innovation, and the firm that delivers a particular innovation first obtains the reward associated with that innovation.

Rewards are connected to other rewards through a small-world network (Fig. 10.4). That is, they are connected to some of their neighbours and a few of their non-neighbours. When two rewards are connected they share one or more knowledge requirements. This property of the knowledge space aims to represent the technical or knowledge connections between inventions used in different areas.

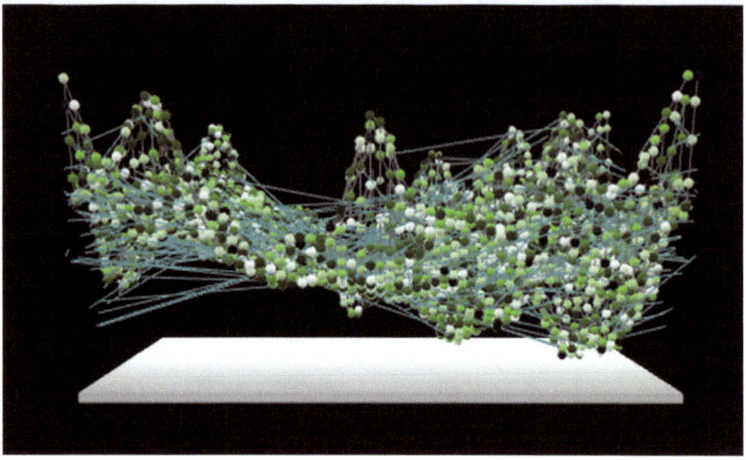

Fig. 10.4 Knowledge links between innovations

After a successful invention, actors may find new uses for some of the knowledge components of the invention. In the model, when a firm fulfils all the knowledge requirements of an innovation, it may capture a financial reward that equals the height (or the z coordinate) of the corresponding reward. In addition, following a successful innovation, some of the actors working at the firm obtain a close-to-perfect vision of a connected reward, which they may use for future projects.

The rewards are semi-visible to the actors. Accordingly, all rewards have a general visibility attribute, and the higher the visibility attribute the darker the colour of a reward in the graphic representation of the model (Fig. 10.5). The level of visibility of a reward for a given actor also depends on the knowledge base of the actor in that field or application area. During the simulation experiments, actors make estimates about the size and knowledge requirements of rewards, and the level of visibility of a reward for an actor determines the accuracy of these estimates. Actors build estimates about a rather invisible reward by drawing numbers from a relatively flat normal distribution (with the true mean), and the estimates about a more visible reward by drawing numbers from a relatively narrow normal distribution (see March 1991).

A certain proportion of rewards (a parameter of the model) are bogus, that is, they appear as any other reward object but they do not provide any financial benefits to the firm, even when the firm satisfies all corresponding knowledge requirements. Bogus rewards are used in the model as one of the representations of the level of uncertainty related to innovative activities in a given industry.

With respect to the construction of the reward space, the height values of different rewards, which represents how much money they involve, is done via diamond-square algorithm. This algorithm is used in computer games for producing realistic looking landscapes. The algorithm first assigns random numbers to each corner of a square shaped space. The midpoints of all four sides of the square are then indicated and these midpoints are assigned the simple average of the numbers at the two nearest corners of the square. The algorithm then refers to the diamond-

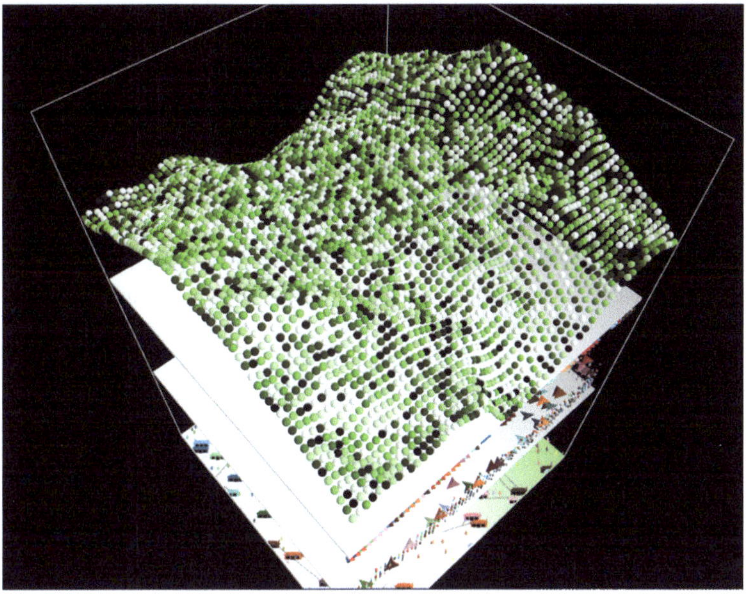

Fig. 10.5 Reward space in three dimensions and semi-visible rewards

shaped area between the indicated midpoints, and like in the previous step, indicates the midpoints of the sides of the diamond and assigns the value equalling the average of their corners to each midpoint. The algorithm continues repeating the same process the number of times specified, and its output is the values associated to the corners and midpoints. This algorithm is used in the AIR model for creating continuity between adjacent rewards, representing multiple innovations of similar importance in a research field and application area.

So far we have introduced the agents and simulation environment of AIR. In the remainder of this section, we will explain the simulated events that occur during our experiments.

The experiments start with the generation of n firms in each region. The firms are randomly assigned an initial capital using a normal distribution with region-level parameter values. A number of exceptionally big firms can also be created through optional parameter settings for experimenting with different industry structures.

The first thing that firms do as part of their setup is establishing a core-team of managers. The size of this team is a function of the initial capital of the firm. The selection of the team is done at two steps. In the first step, firms shortlist unemployed actors with expertise levels that are higher than the population average, and then randomly select actors from this shortlist. Thus, the managers tend to have higher expertise than the researchers (this is not necessarily the case for all manager-researcher pairs).

The managers then explore the reward space trying to find seemingly profitable research projects, and report them back to the firm. Managers explore possible innovations in the areas that they have some knowledge. That is, they explore the parts of the reward space that have the same *x* and *y* coordinates with their kenes. As the knowledge is semi-visible, they can only build a subjective understanding of possible projects. This subjective understanding will be bounded by the visibility attribute of a given reward object, and the level of expertise that the manager has in that knowledge area. Firstly, a manager can only identify the rewards for which the product of the visibility of the reward and the managers relative expertise in that area exceeds a given simulation-level (global) threshold parameter. Secondly, for those rewards that the manager can identify, he or she will guess the magnitude of each knowledge requirement. The guessing process corresponds to a random draw from a normal distribution with the mean equalling the true value of the magnitude of the knowledge requirement and the variance being smaller for managers with high expertise levels and larger for those with lower expertise. Based on these subjective understandings, managers make estimates of project cost and report all projects with positive expected reward minus cost values back to the firm. The firm then selects one of the projects that its managers report back. While this selection can be done in various ways, in the core model, it is assumed that the firm selects the project with the highest expected net return.

Once the project is selected, the firm employs researchers according to the knowledge creation requirements that are thought to be necessary for realising the project. All firms that are looking for researchers concurrently try to employ the actors with highest possible expertise levels. They advertise positions, which are visualised at the knowledge space, and invite applications. In the model it is assumed that researchers prefer to work in their own regions, so unless they remain unemployed for prolonged periods of time they only apply to positions in their own region. After long periods of unemployment, they accept migrating to the other region. New entries of researchers into the model occur regularly. At each period, actors age, those who reach the retirement age exit the model, and new researchers graduating in all knowledge areas enter.

In the model, the research activities within a project proceed in a stochastic way. Learning at different research areas occurs intermittently and in random amounts following a Poisson distribution. In addition, researchers come together with a random subset of their colleagues and friends every period, and they can acquire bits of knowledge during these interactions.

Firms pay salaries to all their employees and this expenditure is deducted from the firm's capital at each period. After starting a project, firms regularly check on the progress of the project. They may decide to employ additional researchers or collaborate with other firms in knowledge creation requirements that they lag most. These attempts correspond to scanning of the knowledge space and identifying researchers and firms that have expertise.

As salaries are paid to the researchers and managers that are employed, firms need to generate income through successful innovation projects. After completion of a project the managers look for another project for the firm, and it continues to operate. If, however, the capital of the firm falls to zero, it bankrupts. This evolutionary setting, where firms with inadequate capital are replaced with new entries is similar to the SKIN model. Yet, due to the multi-level structure of AIR, bankruptcy of a firm does not correspond to total disappearance of its innovative efforts as its employees can work elsewhere, bringing the knowledge they have accumulated during the past projects of the bankrupting firm to their new firms. Hence, AIR makes it possible to take into account the effects of unsuccessful projects and firms in the knowledge base of a region, which are often neglected in existing models.

New entries of firms are modelled through entrepreneurial activities of actors. Here, in line with the reviewed literature, it is conceptualised that regions have an entrepreneurial character, or infrastructure that influences individual tendencies. Therefore in the model, the value of the entrepreneurial attitude of actors is linked with the regional entrepreneurship parameter; the value of the actors' attitude is assigned from a normal distribution with the mean equalling the regional parameter. Hence, if the entrepreneurship parameters of the two simulated regions are set differently, the actors in one region may be on average more willing to set up their own businesses than those in the other region.

Entrepreneurial actions can occur through spin offs from successful projects or independently from existing activities. In the former, after a successful innovation, managers of the innovating firm identify rewards that were linked to the one that was obtained. They then decide whether or not they will establish their own company aiming to achieve these new opportunities. In the latter, a limited number of actors are picked each period, and they are asked to decide whether or not they will establish their own companies. In both types of entrepreneurial decisions, actors draw a random number, and if it is smaller than their entrepreneurial attitude they found their company.

10.3.3 Model Application: Generating Silicon Valley and Route 128 In Silico

In this subsection, we aim to illustrate how the features that AIR model adds on to the SKIN model can be useful for studying clusters and regional innovation networks. Here, we focus on entrepreneurship as the subject of our experiments, and on the two-regions structure of the model to show how it can be used for comparative simulation experiments. This feature is particularly well fitted to study dynamics of two competing regions in the same industry.

In her well-known comparative case study of Silicon Valley and Route 128, Saxenian (1996) points out the similarities in the early histories and structural conditions of the regions, and the divergence in their performances after 1980s. She then relates the valley's success to a number of interrelated factors. She describes, for example, the migration of a generation of young, mostly male experts to the valley, both rejecting the formal, hierarchical organisational structures of the Route 128 firms, and exploiting the possibility of working for interesting and state-of-art projects. Coming to the valley without a family, these young people were open to taking risks and experimentation. Flat and open organisations spinning off from

Fairchild led to the development of a collaborative network that fits the increasingly dynamic nature of the semi-conductor industry. Saxenian also reports differences between the two regions in communicating entrepreneurial success. Whereas, Route 128 firms kept their financial status to themselves, the generous life styles of the successful valley entrepreneurs were visible to the public. Overall, the valley attracted migration of influential actors who seemed to have followed each other.

Sorenson and Audia (2000; p. 442) point out that geographic concentration also increases failure rates. They therefore argue that agglomerations must be driven by even higher entry rates. Industries remain concentrated, they argue, not so much due to the efficiency of particular locations but because concentration increases the pool of potential entrepreneurs "*in a manner similar to a pollination process in which plants produce pollen that blows away in the wind only to land somewhere nearby and burst into new plants*". This is because of changes occurring in the structure of entrepreneurial opportunities. Entrepreneurs learn the trade from existing organizations. Individuals would be presented with more opportunities to access knowledge, and form critical networks and thus the likelihood of them leaving their current employers and becoming entrepreneurs would increase. Looking at the shoe industry, they conclude that it is variation in the structure of entrepreneurial opportunities that maintains geographic concentration in the sector.

The AIR model can be used for history-friendly experiments that generate conditions and processes that can replicate Silicon Valley's success compared to Route 128. Building on Sorenson and Audia's (2000) emphasis on entry rates, and Saxenian's (1996) observations on migration to and entrepreneurship in Silicon Valley, we aimed to experiment if we could generate agglomeration effects only by altering the entrepreneurial parameter of the two regions while keeping everything else constant.

For this purpose, as explained in the previous section we introduced explicit entrepreneurial processes into the AIR model. It may be useful to repeat here that entrepreneurship in AIR happens in two ways. Firstly, each period a number of actors are asked to decide whether or not they will found their own firms; and secondly, after a successful innovation managers of the innovating firm identify other innovation opportunities, and decide whether they will start up their own firms and pursue these opportunities.

Entrepreneurs in the model can either choose to start their new businesses close to where they live and work, for benefiting from existing clientele and supply chains, or they can return to their home regions for making use of personal contacts and access to informal credits. The model is suitable for comparing cumulative effects of such patterns, which may have different weights in different industries. In this chapter, we are keeping our *in silico* history friendly experiments rather simple and keeping these reflections on literature as future research avenues.

After implementing this structure, we run the simulation experiments increasing only the entrepreneurial parameter of one region (Silicon Valley) and keeping everything else the same for the two regions. The model effectively re-generates the comparative success of the Silicon Valley. In the experiment illustrated in Figs. 10.6 and 10.7, the valley ends the experiments with 200 firms, compared to 74 in Route 128.

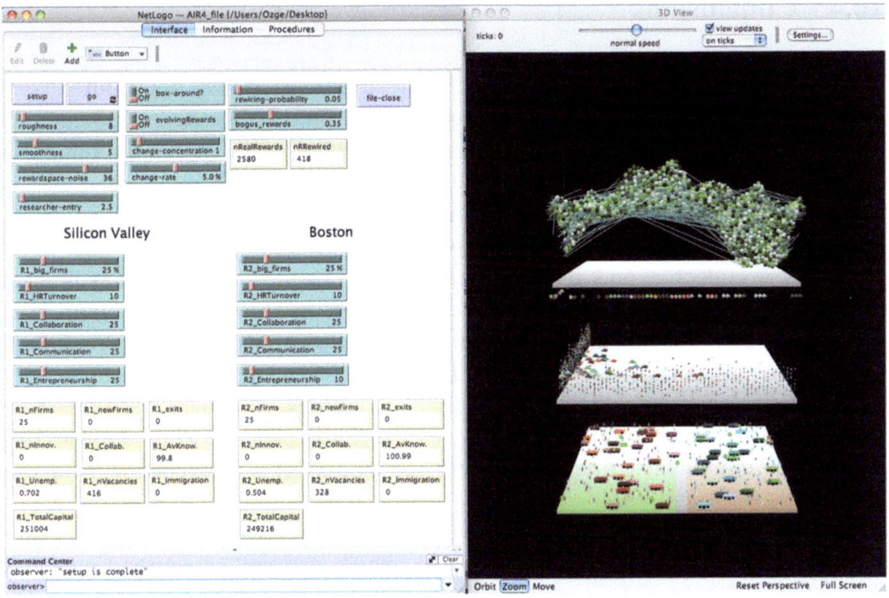

Fig. 10.6 The model at the beginning of an experiment

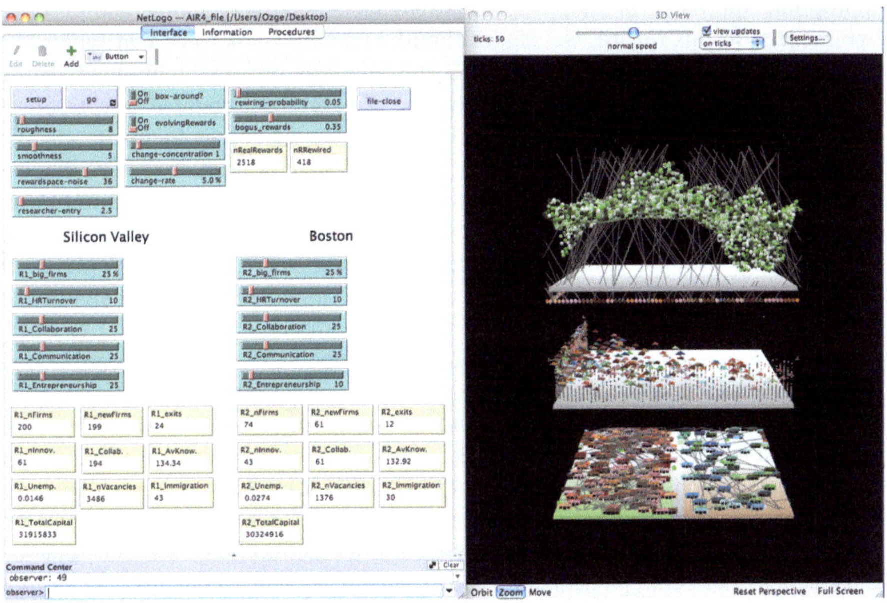

Fig. 10.7 The model at the end of an experiment

10.4 Conclusions

In this chapter, we address evolution of industrial clusters, we point out several dimensions of the complexity of this evolutionary process, we develop modelling ideas that can be used for tackling this complexity and finally introduce our extension of the SKIN model that implements these modelling ideas.

For getting hold of the complexity of cluster emergence, we first reflect on the existing literature on the nature of innovative processes. We describe innovative processes as collective and distributed creation of knowledge and sense-making at the face of inherent uncertainties. We then elaborate on the agency of firm, underlining path-dependent and idiosyncratic processes of knowledge creation as well as the bottom-up emergence of firm-level behaviour through interactions within formal and informal organisational structures. We then position these dynamic and interdependent formations in a regional context, while emphasising the role of individual entrepreneurs and the culture of entrepreneurship in this context.

We suggest agent-based models are suitable research tools for addressing this complexity. We start with the SKIN model, which is a well-established model of innovation networks and extend the model in multiple directions for addressing the complexity of industrial clusters as comprehensively as possible.

Firstly, we suggest that employing a modular development procedure can be an appropriate approach for balancing the strengths of relatively simple and descriptive models. In this regard, we suggest building a core framework that represents the main conceptual spheres related to the research subject highlighted in the literature and then in time, during the course of several interrelated research projects, integrating relatively simple, well-grounded models that focus on specific research questions in this core framework in a step-wise fashion. We argue that the resulting model will have the advantages of conceptual coherence and flexibility for different research purposes, while individual modules remain relatively easy to communicate, discuss, test and replace. Regarding our model on evolution of industrial clusters, we expect that this modelling approach can make significant contributions to the existing literature, linking case studies to macro-level findings.

We also propose an additional level of agency to the SKIN model for representing the role of actors in addition to firms in innovation processes. This additional level adds richness and micro-foundations to our analysis. This more detailed and dynamic representation of the firm also enables us to open up the black box of the firm and study the interactions within the firm. Thus it also fits the evolutionary perspectives of the firm in the innovation studies quite well. In our model implementation, we use the multi-level structure of the model for studying the connections between regional constructions of entrepreneurship and individuals' entrepreneurial behaviour over time.

Finally, we introduce elements of intentionality to the stochastic processes of the SKIN model. In particular, we conceptualise a semi-visible space of possible projects that certain employees of the firm explore. This space is designed to reflect uncertainties inherent in innovative processes, bounded rationality and intentionality of agents within their subjective realities around research and development projects. After the selection of a target project the firm focuses its operations towards completing the project. This new feature provides a more realistic representation of innovative activities of firms and enables us to address strategies and organisational structures of firms more directly.

We illustrate the model and how it can be used with a module focusing on the role of entrepreneurship in the development of clusters. The module is integrated into the core AIR structure. Our illustrative experiments regenerate the success of the Silicon Valley in comparison to Route 128. Accordingly, it is possible to reproduce the geographical agglomeration of firms operating in an industry in a given region, only by increasing the entrepreneurial parameter of one region and keeping everything else constant. Our preliminary results coincide with Sorenson and Audia's (2000) insights that agglomeration is triggered by increased founding rates, despite greater failure rates (exits) caused by increased local competition.

References

Ahrweiler P, Pyka A, Gilbert N (2004) Simulating knowledge dynamics in innovation networks. In: Leombruni R, Richiardi M (eds) Industry and labor dynamics: the agent-based computational economics approach. World Scientific, Singapore, pp 284–296

Anselin L, Varga A, Acs Z (1997) Local geographical spillovers between university research and high technology innovations. J Urban Econ 42(3):422–448

Audretsch DB (1995) Innovation and industry evolution. MIT Press, Cambridge, MA

Audretsch DB, Falck O, Feldman MP, Heblich S (2012) Local entrepreneurship in context. Reg Stud 46(3):379–389

Boero R, Castellani M, Squazzoni F (2004) Micro behavioural attitudes and macro technological adaptation in industrial districts: an agent-based prototype. J Artif Soc Soc Simulat 7(2)

Boschma RA, Frenken K (2009) Some notes on institutions in evolutionary economic geography. Econ Geogr 85(2):151–158

Brenner T (2001) Simulating the evolution of localised industrial clusters – an identification of the basic mechanism. J Artif Soc Soc Simulat 4(3)

Breschi S, Malerba F (2005) Clusters, networks and innovation. Oxford University Press, Oxford

Cooke P, Morgan K (1993) The network paradigm: new departures in corporate and regional development. Environ Plann D Soc Space 11(5):543–564

Cooke P, Gomez Uranga M, Etxebarria G (1997) Regional innovation systems: institutional and organisational dimensions. Res Pol 26(4–5):475–491

Cowan R, Jonard N (2004) Network structure and the diffusion of knowledge. J Econ Dynam Contr 28(8):1557–1575

Cowan R, Jonard N, Zimmermann JB (2005) Bilateral collaboration and emergent networks. Available at SSRN: http://ssrn.com/abstract=882420 or http://dx.doi.org/10.2139/ssrn.882420

Cyert RM, March JG (1963) A behavioral theory of the firm. Prentice Hall, Englewood Cliff, NJ

Doms M, Lewis E, Robb A (2010) Local labor force education, new business characteristics, and firm performance. J Urban Econ 67(1):61–77

Dosi G, Marengo L (1994) Toward a theory of organizational competences. In: England RW (ed) Evolutionary concepts in contemporary economics. Michigan University Press, Ann Arbor, MI, pp 157–178

Feldman MP (2001) The entrepreneurial event revisited: firm formation in a regional context. Ind Corp Change 10(4):861–891
Feldman MP, Francis J, Bercovitz J (2005) Creating a cluster while building a firm: entrepreneurs and the formation of industrial clusters. Reg Stud 39(1):129–141
Fioretti G (2001) Information structure and behaviour of a textile industrial district. J Artif Soc Soc Simulat 4(4):1–30
Florida R (2005) Cities and the creative class. Routledge, New York
Foster J (2005) From simplistic to complex systems in economics. Camb J Econ 29(6):873–892
Glückler J (2007) Economic geography and the evolution of networks. J Econ Geogr 7(5):619–634
Hodgson GM (1998) Competence and contract in the theory of the firm. J Econ Behav Organ 35 (2):179–201
Hofstede G (2001) Culture's consequences – comparing values, behaviours, institutions and organizations across nations. Sage, London
Holcombe RG (2003) The origins of entrepreneurial opportunities. Rev Austrian Econ 16(1):25–43
Holland JH (1975) Adaptation in natural and artificial systems: an introductory analysis with applications to biology, control, and artificial intelligence. University of Michigan, Ann Arbor, MI
Howells J (1999) Regional systems of innovation? In: Archibugi D, Howells J, Michie J (eds) Innovation policy in a global economy. Cambridge University Press, Cambridge, pp 67–93
Jalonen H (2012) The uncertainty of innovation: a systematic review of the literature. J Manag Res 4(1):1–47
Koschatzky K, Sternberg R (2000) R&D cooperation in innovation systems—some lessons from the European Regional Innovation Survey (ERIS). Eur Plann Stud 8(4):487–501
Krugman P (1991) Geography and trade. MIT Press, Cambridge, MA
Landier A (2001) Entrepreneurship and the stigma of failure. MIT Job Market Paper
Lane DA, Maxfield RR (2005) Ontological uncertainty and innovation. J Evol Econ 15(1):3–50
Malerba F, Orsenigo L (2000) Knowledge, innovative activities and industrial evolution. Ind Corp Change 9(2):289–314
March JG (1991) Exploration and exploitation in organizational learning. Organ Sci 2(1):71–87
Marengo L (1992) Coordination and organizational learning in the firm. J Evol Econ 2(4):313–326
Marengo L (1996) Structure, competence and learning in an adaptive model of the firm. In: Dosi G, Malerba F (eds) Organization and strategy in the evolution of the enterprise. Macmillan, London, pp 124–154
Marshall A (1920) Principles of economics: an introductory volume. Macmillan, London
Maskell P, Malmberg A (2007) Myopia, knowledge development and cluster evolution. J Econ Geogr 7(5):603–618
Mises L (1998 [1949]) Human action: a treatise on economics. Ludwig von Mises Institute, Auburn, AL
Nelson R (1991) Why do firms differ, and how does it matter? Strat Manag J 14(S2):61–74
Nonaka I, Tomaya R, Nagata A (2000) A firm as a knowledge-creating entity: a new perspective on the theory of the firm. Ind Corp Change 9(1):1–20
Otter HS, Veen A, de Vriend HJ (2001) Location behaviour, spatial patterns, and agent-based modelling. J Artif Soc Soc Simulat 4(4)
Prahalad C, Hamel G (1990) The core competence of the corporation. Harv Bus Rev 68(3):79–91
Pyka A (2007) Innovation networks. In: Hanusch H, Pyka A (eds) Elgar companion to neo-Schumpeterian economics. Edward Elgar, Cheltenham
Pyke F, Becattini G, Sengenberger W (eds) (1990) Industrial districts and inter-firm cooperation in Italy. International Institute for Labour Studies, Geneva
Rosenberg N (2003) Innovation and economic growth. In: OECD conference on innovation and growth in tourism, Lugano, Switzerland, 18–19 Sept 2003, manuscript available at http://www.oecd.org/cfe/tourism/34267902.pdf

Saxenian AL (1996) Regional advantage: culture and competition in Silicon Valley and Route 128. Harvard University Press, Cambridge, MA

Schumpeter JA (1934 [1912]) The theory of economic development. Harvard University Press, Cambridge, MA

Schumpeter JA (1939) Business cycles: a theoretical, historical, and statistical analysis of the capitalist process. McGraw-Hill, New York

Simon H (1976) From substantive to procedural rationality. In: Latsis SJ (ed) Method and appraisal in economics. Cambridge University Press, Cambridge

Sorenson O, Audia PG (2000) The social structure of entrepreneurial activity: geographic concentration of footwear production in the United States, 1940–1989. Am J Sociol 106(2):424–462

Stam E (2010) Entrepreneurship, evolution and geography. In: Boschma R, Martin R (eds) The handbook of evolutionary economic geography. Edward Elgar, London, pp 307–348

Sternberg R (2009) Regional dimensions of entrepreneurship. Found Trends Entrepren 5(4):211–340

Teece D, Pisano G, Shuen A (1997) Dynamic capabilities and strategic management. Strat Manag J 18(7):509–533

Watts DJ, Strogatz SH (1998) Collective dynamics of ‘small-world’ networks. Nature 393 (6684):409–410

Zhang J (2003) Growing silicon valley on a landscape: an agent-based approach to high-tech industrial clusters. J Evol Econ 13(5):529–548

Index

N. Gilbert et al. (eds.), *Simulating Knowledge Dynamics in Innovation Networks*, Understanding Complex Systems, DOI 10.1007/978-3-662-43508-3,

MIX
Papier aus verantwortungsvollen Quellen
Paper from responsible sources
FSC® C105338

FSC
www.fsc.org

If you have any concerns about our products,
you can contact us on
ProductSafety@springernature.com

In case Publisher is established outside the EU,
the EU authorized representative is:
Springer Nature Customer Service Center GmbH
Europaplatz 3, 69115 Heidelberg, Germany

Printed by Libri Plureos GmbH
in Hamburg, Germany